PHARMACY
INFORMATICS

PHARMACY INFORMATICS

Philip O. Anderson

Susan M. McGuinness

Philip E. Bourne

CRC Press
Taylor & Francis Group
Boca Raton   London   New York

CRC Press is an imprint of the
Taylor & Francis Group an **informa** business

A CHAPMAN & HALL BOOK

Back cover image by Erik Matwijkow.

CRC Press
Taylor & Francis Group
6000 Broken Sound Parkway NW, Suite 300
Boca Raton, FL 33487-2742

Printed in the United States of America on acid-free paper
10 9 8 7 6 5 4 3 2 1

International Standard Book Number: 978-1-4200-7175-7 (Hardback)

Library of Congress Cataloging-in-Publication Data

Pharmacy informatics / editors: Philip O. Anderson, Susan M. McGuinness, Philip E. Bourne.
   p. ; cm.
  Includes bibliographical references and index.
  ISBN 978-1-4200-7175-7 (hardcover : alk. paper)
  1. Pharmacy--Information services. 2. Pharmacy--Data processing. 3. Medical informatics. I. Anderson, Philip O. II. McGuinness, Susan M. III. Bourne, Philip E. IV. Title.
  [DNLM: 1. Clinical Pharmacy Information Systems. 2. Drug Information Services. 3. Information Storage and Retrieval. 4. Medical Records Systems, Computerized. QV 26.5 P5356 2010]

RS56.2.P48 2010
615'.1--dc22
                                        2009036641

Visit the Taylor & Francis Web site at
http://www.taylorandfrancis.com

and the CRC Press Web site at
http://www.crcpress.com

Contents

Section III **Information Systems**

Section IV **Decision Support**

SECTION V **The Future of Pharmacy Informatics**

Foreword

Sequencing of the human genome did not simply carry us into a postgenomic era; rather, it was a critical milestone in the start of genomic perspectives that will affect how we approach disease, diagnosis, and therapy. Similarly, the emergence of genomic information was a watershed development that led to the field of bioinformatics and will influence pharmacy practices and pharmaceutical research and development for the foreseeable future.

We have reached new levels in the information age in which contemporary informatics will enable us to collect and handle the explosion of data that can be mined from therapeutic outcomes. Personalized medicine and pharmaceutical care, with linkages to genomic information, will not begin to serve society optimally until they translate to the practitioners positioned to apply them to large segments of the world population. Sophisticated quantitative approaches to data handling and assessment are essential to future advances. Of equal importance, however, is the education of the practitioner who will be engaged in individualizing therapy and communicating outcomes to the patient.

Practicing pharmacists and physicians will be expected to use databases to ensure medication efficacy, patient safety, and confidentiality. Telepharmacy and telemedicine with patient information networks will become more common, and pharmaceutical research and development will assume more global positions. Such endeavors require new means of information transfer for which our academic institutions should be taking a leadership position.

Continual growth and increasing complexity of therapeutic information require our colleges and schools of pharmacy to prepare the practitioner not only for practice after licensure but also for a practice that can respond to the scientific advances over the four decades of an active career. In starting a new pharmacy school on a research-intensive campus with a burgeoning internal healthcare system, members of the faculty at the University of California, San Diego (UCSD), have attempted to address this issue. We feel that informatics not only will play a greatly expanded role in pharmacy education, but also will undergo substantive technological advances.

To develop a core course in informatics and apply its principles throughout the curriculum, we have asked faculty with expertise in drug information, library sciences, bioinformatics, and the computational technologies of high-performance computing to lead this endeavor. In turn, they have engaged a variety of faculty in different practice settings to consider applications as contributing authors. Curricular and textbook design encompasses what a traditional faculty might assemble, but is also informed by individuals outside pharmacy circles.

Doctors Phil Anderson, Sue McGuinness, and Phil Bourne have contributed to and edited this reference informatics textbook for the student pharmacists in their course. This course follows a prior introductory course that exposes the student to computer skills and biostatistics. The pharmacy informatics course is positioned to offer the basic principles for courses in study design, therapeutics, drug information, and pharmacogenomics delivered later in the curriculum. The diversity in backgrounds of the three editors has enabled them to fashion a textbook that extends beyond the UCSD curriculum and should serve as a treatise for the evolving curricula in pharmacy to refine over the years.

Palmer Taylor, Ph.D.
Dean, Skaggs School of Pharmacy and Pharmaceutical Sciences
Sandra and Monroe Trout Chair in Pharmacology
Associate Vice-Chancellor for Health Sciences
University of California, San Diego

Preface

Pharmacy practice, like all areas of healthcare delivery, is in a state of rapid change—some would say crisis. These changes are being driven by an aging population, new legislative initiatives, healthcare costs, the promise of genomic medicine, expanding roles for pharmacists in healthcare, and upheavals in the pharmaceutical industry that include company mergers, increased emphasis on biologicals, limited pipelines of new drugs, and drug recalls. Information technology is seen as both a driver of these changes and a response to the perceived crisis. It is timely, therefore, to introduce a book specifically on the subject of pharmacy informatics.

Taken separately, "pharmacy" and "informatics" are broad subjects. For the past 7 years, we have grappled with offering the appropriately focused doctor of pharmacy course that provides the training needed to cope with changes in pharmacy practice—whether future careers will be in community pharmacy, hospital pharmacy, the pharmaceutical industry, or other healthcare sectors. This book is a result of that process. Although it is intended as a textbook for our course, we hope that it appeals not only to the pharmacists of tomorrow, but also to pharmacists already in practice and interested consumers of pharmacy services.

We are part of the new and dynamic Skaggs School of Pharmacy and Pharmaceutical Sciences (SSPPS) at the University of California, San Diego. The doctor of pharmacy and doctor of pharmacy/doctor of philosophy programs reflect our unique association with the School of Medicine: Students undergo part of their training with medical students and the broader health sciences campus, as well as having access to UCSD's strengths in the biomedical and computer sciences. We are also fortunate that healthcare informatics is quite advanced in San Diego and that we are able to tap experts from other local institutions for our course and this textbook. We hope that our excitement in being part of a new enterprise, unafraid of pushing the boundaries while still providing all the fundamental elements of more traditional training, comes across in this book.

Given the breadth of topics that fall under pharmacy informatics, no one or two individuals could hope to provide the expertise needed in all areas. We have called upon all our course lecturers to participate in this work and, just as we seek continuity and relevance in the course material, we have tried to organize the book in the same way. Pharmacy informatics is learned by doing. Our course reflects this principle. Each class of two 3-hour sessions per week for 10 weeks consists of a prerequisite presentation of background material by the lecturer, followed by hands-on exercises.

It is fitting that we start with a foreword by the dean of our new school, Dr. Palmer Taylor, because his vision, more than anything else, saw the inclusion of this course as part of the core curriculum for all doctor of pharmacy students. The book is divided into five parts, each of which builds upon what has been discussed before. We hope that this organization will facilitate the use of this book as a textbook and also as a reference source where appropriate discussion can be found.

- Section I introduces the scope of the material to be covered and the motivators for the book. Drs. Susan McGuinness, pharmacy librarian, Philip Anderson, coordinator of UCSD's drug information course, and Philip Bourne, bioinformatics expert define what we mean by pharmacy informatics. This overview is followed by a discussion of information and biomedical technologies that are the drivers of change by Howard Asher, president and CEO of Global Life Sciences, Inc., and Philip Bourne, professor of pharmacology at UCSD.

- Section II provides the prerequisites for the effective use of the informatics resources discussed subsequently. Dr. Bourne discusses the basics of maintaining the reliability and security of computers in a connected world. Among both practicing pharmacists and our students, basic knowledge of computing varies widely and we have organized the material so that sections can be easily skipped. A theme that appears repeatedly is the need for standardization in the healthcare industry, including the information contained therein. Doctors McGuinness and Bourne discuss the standards and controlled vocabularies that pharmacists will confront throughout their careers. Dr. McGuinness discusses effective strategies for navigating, searching, evaluating, and managing the wide variety of information resources available today.

- Section III covers the types of information systems that exist in hospitals and pharmacies. The electronic health record (EHR) underlies all these systems and is discussed by Dr. Joshua Lee, associate clinical professor of medicine, who practices both clinically and in informatics at UCSD. Doctors Daniel Boggie, Jennifer Howard, and Armen Simonian, informatics pharmacists from neighboring affiliated healthcare institutions, review the basic elements of pharmacy information and automation systems. Bar coding, a relatively new technology for reducing medication errors, is described by Dr. Ashley Dalton, who was instrumental in implementing this technology at UCSD Medical Center. Dr. Simonian returns to describe how pharmacists and students interested in a pharmacy informatics career can prepare for and function in this role.

A critical aspect of any of these systems is the need to avoid errors. Dr. Joseph Scherger, a professor of family and preventive medicine, and Dr. Grace Kuo, an associate professor of clinical pharmacy in our school, elaborate on medical errors and review how information technology can reduce errors in healthcare delivery. Drug information systems and Web-based resources are vital tools for pharmacists and their effective use is discussed by Doctors Anderson and McGuinness. Finally, Dr.

Joseph Ennesser, a graduate of the founding class of the Skaggs School of Pharmacy and Pharmaceutical Sciences and now a practicing pharmacist, describes the use of personal digital assistants (PDAs) and their roles in today's pharmacy practice.

- Section IV details the next step, where systems are used beyond the basic recall of information to help in decision support of patients. Dr. Laura Nicholson, a hospitalist from a neighboring institution, Scripps Health, has particular expertise in evidence-based medicine and discusses tools for use in evidence-based practice. Dr. Anderson returns with a review of computerized clinical pharmacokinetics methods as a decision support tool. Dr. Pieter Helmons, a pharmacoeconomics specialist at UCSD Medical Center with extensive experience in clinical decision support, elaborates on this promising technology. Dr. Robert Schoenhaus, a pharmacist specializing in pharmacoeconomics, discusses ways in which data contained in information systems can be mined to improve therapy, decrease adverse outcomes, and cut costs.

- Section V takes us to the future of pharmacy informatics and what will drive the field. Dr. Bourne discusses the various developments driven by the Internet, such as the emergence of virtual communities, video on demand, and changing publishing models. Dr. Richard Peters provides the perspective of a pioneer in advanced health-care informatics on how current informatics solutions have developed and how they need to evolve to maximize their potential.

Taken together, the five sections of this book reflect a changing pharmacy profession in which information plays a central role if that practice is to be conducted in the most productive and efficient way for both producer and consumer of healthcare services. We have tried to capture what such a change means to the pharmacy student and the practicing pharmacist, as well as to prepare the reader for what lies ahead in a world characterized by only one certainty: It will be very different.

Philip O. Anderson
Susan M. McGuinness
Philip E. Bourne

Editors

Philip O. Anderson is Health Sciences Clinical Professor of Pharmacy at the University of California, San Diego, Skaggs School of Pharmacy and Pharmaceutical Sciences, where he is in charge of the drug information course. Dr. Anderson is also a member of the coordinating faculty team for the pharmacy informatics and therapeutics courses and lecturer in the pharmacy practice course. He is also a partner in Healthware, Inc., where he helped develop T.D.M.S. 2000, and is the author of the National Library of Medicine's LactMed database.

Susan M. McGuinness is the pharmacy librarian at the Biomedical Library at the University of California, San Diego, and Assistant Clinical Professor at the Skaggs School of Pharmacy and Pharmaceutical Sciences. Dr. McGuinness is chair (2009–2010) of the Libraries and Educational Resources Section of the American Association of Colleges of Pharmacy. She is a member of the coordinating faculty team for the pharmacy informatics course and lectures in the drug information, pharmacy practice, pharmaceutical chemistry, and pharmaceutics courses.

Philip E. Bourne is Professor of Pharmacology at the University of California, San Diego, Skaggs School of Pharmacy and Pharmaceutical Sciences, where he is the lead faculty member for the pharmacy informatics course. Dr. Bourne is also currently the editor-in-chief of *PLoS Computational Biology* and associate director of the RCSB Protein Data Bank, a vital public resource used in drug discovery.

Contributors

Philip O. Anderson, Pharm.D., FCSHP, FASHP
Health Sciences Clinical Professor
Skaggs School of Pharmacy and
    Pharmaceutical Sciences
University of California San Diego
La Jolla, California

Howard R. Asher
President and CEO
Global Life Sciences, Inc.
San Diego, California

Daniel T. Boggie, Pharm.D.
Director, Pharmacy Data Applications
Veterans Administration San Diego
    Healthcare System
San Diego, California

Philip E. Bourne, Ph.D., FAMIA
Professor of Pharmacology
Skaggs School of Pharmacy and
    Pharmaceutical Sciences
University of California San Diego
La Jolla, California

Ashley J. Dalton, Pharm.D.
Clinical Pharmacist
University of California San Diego Medical
    Center—Thornton Hospital
La Jolla, California

Joseph J. Ennesser, Pharm.D.
Pharmacy Business Partner
Target Corporation
Riverside, California

Pieter J. Helmons, Pharm.D.
Pharmacoeconomics Specialist
University of California San Diego Medical
    Center—Hillcrest
San Diego, California

Jennifer J. Howard, Pharm.D.
Director, Pharmaceutical Integrated
    Technologies
Veterans Administration San Diego
    Healthcare System
San Diego, California

Grace M. Kuo, Pharm.D., M.P.H.
Associate Professor of Clinical Pharmacy
Skaggs School of Pharmacy and
    Pharmaceutical Sciences
Associate Adjunct Professor of Family and
    Preventive Medicine
School of Medicine
University of California, San Diego
La Jolla, California

Joshua Lee, M.D.
Associate Clinical Professor of Medicine
University of California San Diego Medical
    Center—Hillcrest
San Diego, California

Susan M. McGuinness, Ph.D., M.L.S.
Assistant Clinical Professor
Skaggs School of Pharmacy and
  Pharmaceutical Sciences
Pharmacy Librarian, Biomedical Library
University of California San Diego
La Jolla, California

Laura J. Nicholson, M.D., Ph.D.
Health Sciences Associate Clinical
  Professor
University of California San Diego School
  of Medicine
Faculty in Graduate Medical Education
Scripps Clinic Medical Group
La Jolla, California

Richard M. Peters Jr, M.D.
Emergency Physician
Southern California Permanente Medical
  Group
Independent Health IT Consultant
La Jolla, California

Joseph E. Scherger, M.D., M.P.H.
Clinical Professor of Family and Preventive
  Medicine
School of Medicine
University of California San Diego
San Diego, California

Robert H. Schoenhaus, Pharm.D.
Pharmacy Benefits Administration
Sharp HealthCare
San Diego, California

**Armen I. Simonian, Pharm.D., FCSHP,
FASHP**
Pharmacy Informatics Specialist
Sharp HealthCare
San Diego, California

Palmer Taylor, Ph.D.
Sandra & Monroe Trout Chair in
  Pharmacology
Dean, Skaggs School of Pharmacy &
  Pharmaceutical Sciences
Associate Vice Chancellor Health Sciences
University of California San Diego
La Jolla, California

I

Introduction

CHAPTER 1

What Is Pharmacy Informatics?

Philip O. Anderson, Susan M. McGuinness, and Philip E. Bourne

CONTENTS

P HARMACISTS PRACTICING TODAY IN the United States or other developed or developing countries will interact with technology in almost every aspect of their work. Government initiatives are driving healthcare systems toward the adoption of health information technology with the goal of higher quality, more cost-effective patient care.[1] At the University of California, San Diego (UCSD), Skaggs School of Pharmacy and Pharmaceutical Sciences (SSPPS), this recognition led to the integration of pharmacy informatics into the doctor of pharmacy curriculum in 2001.

Informatics affects all three curricular areas of emphasis: basic pharmaceutical sciences (e.g., pharmacogenomics, pharmaceutical chemistry, pharmacokinetics), pharmaceutical technology and management (e.g., pharmaceutics, pharmacoeconomics, study design), and clinical pharmacy practice (e.g., therapeutics, drug information, diagnostics, patient counseling). This textbook summarizes the content of a pharmacy informatics course developed by the editors. As such, it reflects a number of years of experience in defining what is relevant to teach in a course that meets the needs of the curriculum, with emphasis on the mission of SSPPS and the changing landscape of healthcare.

Informatics is the study of the best practices in information accrual, handling, dissemination, and comprehension using appropriate technology. Pharmacy informatics deals with the subset of informatics relevant to the practice of pharmacy. As such, it intersects with two other subdisciplines of information science that have a longer history.

First, medical informatics came about as the healthcare system increasingly relied on information technology for data management, communications, decision support, and, ultimately, improved patient care. The American Medical Informatics Association (AMIA) defines medical informatics as "the use of information science and technology to advance medical knowledge and improve quality of care and health system performance." The American Association of Medical Colleges defines it as "the rapidly developing scientific field that deals with resources, devices, and formalized methods for optimizing the storage retrieval and management of biomedical information for problem solving and decision making."

Later came bioinformatics, a field that sprang from the human genome project and used computers to analyze and interpret the vast amounts of data generated in the field of biology. *Stedman's Medical Dictionary* defines bioinformatics as "the scientific discipline encompassing all aspects of biologic information acquisition, processing, storage, distribution, analysis, and interpretation that combines the tools and technology of mathematics, computer science, and biology with the aim of understanding the biologic significance of a variety of data."

Putting it all together, the American Society of Health-System Pharmacists (ASHP) defines pharmacy informatics as an important subset of medical informatics in which pharmacists "use their knowledge of information systems and medication-use processes to improve patient care by ensuring that new technologies lead to safer and more effective medication use."[2] A recent ASHP survey of U.S. hospitals found that information technologies are used in all steps of medication-use processes and that pharmacists must understand these technologies.[3]

"Pharmacy informatics" has been defined in some contexts as the pharmacy specialty dealing with pharmacy computerization. Specialized residencies and standards exist for training a cadre of pharmacy informatics specialists.[4] Although this definition of pharmacy informatics is useful, the UCSD course and this textbook are designed primarily for pharmacist generalists and take a more expansive and general view of pharmacy informatics. Practicing pharmacists need to know not only the computer system in use where they work, but also how it relates to larger systems such as the hospital computer system and peripheral automation that is connected to these systems.

Our working model for the pharmacy informatics course is based on the idea that pharmacists must use information and technology skills to integrate information about drugs and information about patient-related issues from a variety of sources in order to achieve patient-centered care (see Figure 1.1).

Information about drugs includes primary literature and electronic information resources, hospital and pharmacy information systems, pharmacokinetics, and pharmacogenomics. Patient-related issues include medication safety, electronic health records, decision support systems, and the practice of evidence-based medicine. The skills needed to implement this knowledge in patient-centered care include effective literature and Web search skills, an understanding of databases, and the controlled vocabularies needed for interoperability between systems and for optimal searching of some databases. Additionally, pharmacists need to know how to access computerized medical information in various databases and to understand the underpinnings of these databases to use them most effectively.

Pharmacy students also need to know the current state of the art in hospital and pharmacy information and decision support systems, how to use these systems, and the components and functions needed to build efficient and effective systems. All of these information and technology skills should help the pharmacist clearly see the complex picture of medication management for each individual patient and make the best possible decisions for his

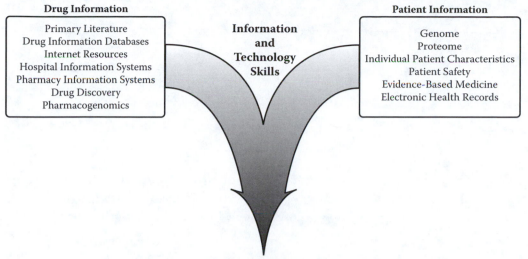

Drug Information

Primary Literature
Drug Information Databases
Internet Resources
Hospital Information Systems
Pharmacy Information Systems
Drug Discovery
Pharmacogenomics

Information and Technology Skills

Patient Information

Genome
Proteome
Individual Patient Characteristics
Patient Safety
Evidence-Based Medicine
Electronic Health Records

Patient-Centered Care

FIGURE 1.1   Working model for pharmacy informatics.

or her care. This textbook covers prerequisite information and technology skills, information systems currently used in hospitals and pharmacies, what a career in pharmacy informatics involves, and what the future of pharmacy informatics might look like. We hope that others find this information helpful and relevant.

REFERENCES

1. United States Office of the National Coordinator for Health Information Technology. Health information technology (http://healthit.hhs.gov/portal/server.pt).
2. American Society of Health-Systems Pharmacists. ASHP statement on the pharmacist's role in informatics. *American Journal of Health-System Pharmacy* 2007. 64:200–203.
3. Pedersen, C. A., and Gumpper, K. F. ASHP National Survey on Informatics: Assessment of the adoption and use of pharmacy informatics in U.S. hospitals—2007. *American Journal of Health-System Pharmacy* 2008. 65:2244–2264.
4. American Society of Health-Systems Pharmacists. PGY2 informatics residency progressive detail lists of educational outcomes, goals, objectives, instructional objectives (http://www.ashp.org/s_ashp/docs/files/RTP_PGY2InforProgressiveDetailLists.doc).

CHAPTER 2

Drivers of Change

Emergent Information and Biotechnologies

Howard R. Asher and Philip E. Bourne

CONTENTS

2.1 INTRODUCTION

Pharmacy and pharmaceutical sciences have been affected, and will be more so in the future, by information technologies and biotechnologies. This book details some of the outcomes of that impact and how they affect pharmacists' professional lives. This chapter introduces some of the elements that comprise these technologies as well as their implications. What is apparent is that these technologies represent drivers of change in a healthcare industry that is considered a late adopter with a low tolerance for risk, particularly with respect to information technology. However, we now speak of this era of digital medicine as if these technologies are about to precipitate major change. Some would argue that we are now, to use a phrase popularized by Malcolm Gladwell, at a "tipping point."[1]

As we describe the gains that can be made by greater adoption of information and biotechnologies against the backdrop of the current state of our healthcare industry (with particular reference to the United States), it is easy to believe that we are at this tipping point. By one estimate by the Rand Corporation, if 90% of U.S. hospitals and physicians were to adopt hospital information systems over the next 15 years, the industry would save $77 billion per year from efficiency gains.[2] If health and safety gains are considered also, these savings could double to 6% of the $2.6 trillion estimated to have been spent on healthcare in 2009. These savings are compelling and it is not surprising that governments are attempting to control escalating healthcare costs through the adoption of better information and biotechnologies. The bottom line is that these changes will have an impact on pharmacists and pharmaceutical scientists because the current system of healthcare is simply not sustainable.

What is the current state of healthcare? What are these drivers of change? How will changes affect healthcare and the pharmacists that provide that care? These are some of the questions addressed in this chapter.

2.2 THE CURRENT SITUATION

We begin by briefly summarizing the state of healthcare in the United States and the state of the drug industry at large to emphasize the scale of the current problems.

2.2.1 The Current State of Healthcare in the United States

The following data on the state of healthcare in the United States surely must be drivers of change because, as was stated previously, the current system is simply not sustainable:

- The $2.6 trillion the United States will spend on healthcare this year represents 17.6% of the U.S. economy; if unchecked, this percentage will rise.[3]

- Of the total money spent on healthcare worldwide, the United States spends 54%.

- Compared with five other developed nations—Australia, Canada, Germany, New Zealand, and the United Kingdom—the U.S. healthcare system ranks last or next to last on quality, access, efficiency, equity, and healthy lives—five dimensions of a high-performance health system. The United States is the only country of the five without universal health insurance coverage; this partly accounts for its poor performance on access, equity, and health outcomes. The inclusion of physician survey data also shows the United States lagging in adoption of information technology and use of nurses to improve care coordination for the chronically ill.[4]

- Overall, the United States ranks 37 out of 191 countries in the quality and performance of healthcare (see Table 2.1).[5]

- The United States ranks 30th in life expectancy.[6]

TABLE 2.1   World Health System Rankings[7]

1. France	11. Norway	21. Belgium	31. Finland
2. Italy	12. Portugal	22. Colombia	32. Australia
3. San Marino	13. Monaco	23. Sweden	33. Chile
4. Andorra	14. Greece	24. Cyprus	34. Denmark
5. Malta	15. Iceland	25. Germany	35. Dominica
6. Singapore	16. Luxembourg	26. Saudi Arabia	36. Costa Rica
7. Spain	17. Netherlands	27. United Arab Emirates	37. *United States*
8. Oman	18. United Kingdom	28. Israel	38. Slovenia
9. Austria	19. Ireland	29. Morocco	39. Cuba
10. Japan	20. Switzerland	30. Canada	40. Brunei

Source: World Health Organization (www.who.int/whr/2000/media_centre/press_release/en/index.html).

- Of each dollar spent on healthcare, 10 cents goes toward medical liability and defensive medicine.

- An estimated 60 million people in the United States have no health insurance.

These statistics represent enough woe as to the state of healthcare and must be incentives to change. Let us now look at the state of drug discovery as another issue that will affect healthcare, including pharmacy practice.

2.2.2 The Current State of Drug Discovery

The following points are taken from the 2009 Outlook Report from the Tufts Center for the Study of Drug Development[8]:

- Through a concerted effort at the Food and Drug Administration (FDA), the time to approve a new drug has dropped in recent years, but seems to have stabilized at 8 years. Drugs that are developed are most often used to treat complex diseases and are not necessarily that effective.

- The cost of bringing a drug to market can be US$1 billion.

- New drug output has stagnated; fewer than 30 drugs were approved in 2007.

- The introduction of therapeutic monoclonal antibodies is increasing and having a positive impact on the rate of drug discovery.

2.3 HISTORICAL EXAMPLES OF PRECIPITATORS OF CHANGE

The preceding facts sound like doom and gloom; although it is fine to say that this will drive the United States to change, is that possible? One way to answer that question is to consider how change has been wrought in the past. Here are a few examples in chronological order:

- Stethoscope (1816). Rene Laennec of France invented the first stethoscope to protect the modesty of one of his female patients. In 1837, Dr. Oliver Wendell Holmes returned from medical studies in Paris and urged his fellow American physicians to increase their use of stethoscopes. By the mid-1840s, the stethoscope had become integral to the practice of medicine in the United States.

- Thermometer (1867). Sir Thomas Allbutt introduced the first thermometer meant to take the temperature of a person.

- X-rays (1895). Wilhelm Conrad Röntgen accidentally discovered x-rays upon seeing an image cast from his cathode ray generator. The announcement of Röntgen's discovery was illustrated with an x-ray photograph of his wife's hand. The x-ray became one of the defining technological devices to move the art of medical diagnosis to a scientifically based medicine in the early 1900s.

- Blood pressure cuff (1901). Harvey Cushing introduced a version of the modern blood pressure cuff (sphygmomanometer) to U.S. physicians.

- Penicillin (1929). Sir Alexander Fleming's discovery of penicillin in 1929 went undeveloped until the 1940s, when Howard Florey and Ernst Chain isolated the active ingredient from *Penicillium* mold and developed a powdery form of the medicine. Under the pressure of World War II, pharmaceutical manufacturers rapidly adopted mass production methods, reducing the production costs to 1/1000th of the original.

Interestingly, these innovations, which we take for granted in today's provision of healthcare, share similar characteristics: an extended period before wide adoption was seen. It may be that even in the accelerated pace of a modern healthcare world, there will be a marked lag time before the technologies introduced subsequently become commonplace. First, we have to reach the tipping point. Assuming these changes do come eventually, which of them will have an impact on pharmacy practice?

2.4 CHANGES EXPECTED TO RESULT FROM INFORMATION TECHNOLOGY

Information technology (IT) remains underused in healthcare. This fact is surprising given that providing adequate healthcare involves managing and effectively using information. It is not that the need for information has not been recognized. For example, the American Medical Informatics Association (AMIA; www.amia.org) has existed for over 30 years and has over 4,000 members. So, why has the uptake of IT within healthcare been slow? In the 1970s, information technology was expensive and alien to most healthcare providers. Centralized mainframe computers provided billing services, but little else.

The emergence of so-called minicomputers saw a diversification of use in a distributed model of computational operation. Thus, for example, the radiology department began using image processing, and various departments began developing and using databases for diverse information ranging from patient records to pathology samples to the tumor registry. These systems required expert personnel and in no way communicated or interoperated with each other.

The late 1970s and early 1980s saw the emergence of intranets: internal computational networks that started to allow these computers to communicate. This was certainly one of the early drivers of medical informatics because it was soon realized that common naming conventions for items of information needed to be used if the information located on these respective computers were to be used collectively (see Chapter 4). In the early 1980s, IT slowly migrated into back office, to inventory control, central supplies, as well as management of pharmaceuticals and other prescribed medical products. The early 1980s also saw the emergence of the personal computer (PC) and a real opportunity to distribute healthcare information. It seems strange now, but only a small fraction of people were adept at using the PC at that time and, in general, healthcare providers were resistant. Further, the cost per PC station was approximately 10–100 times what it is today.

The 1990s changed all that (as further elaborated in Chapter 3). With computer power doubling for the same cost every 18 months and the advent of widespread Internet use by both patients and providers, the stage was set for change. In the early 1990s, IT began a slow adoption within prescription management, processing, prescription label generation, pharmacy billing, and work flow. In the mid-1990s, the Internet began to be recognized as an expedient source of some medical and pharmaceutical information. The late 1990s saw an important pharmacy innovation from a small company in San Diego, California. Pyxis introduced products for automated and controlled medication dispensing and pharmaceutical supply management.

Still, the human factor persists and should not be underestimated in the adoption of any technology. Often people are happy with the old way of doing things and do not see the more institutional (and often global) implications of their inertia. Institutional mandates come into play here. For example, the insistence that the U.S. Veterans Administration hospitals adopt a single, universal system could not be resisted by care providers if they wanted to keep their jobs.[9] Such systems have a sufficiently successful track record and the problems of global health are so pressing that more rapid adoption of IT in all healthcare sectors seems inevitable.

2.4.1 Electronic Health Record

The electronic health record (EHR) is perhaps the single most important component of medical and pharmacy informatics and is discussed in detail in Chapters 6 and 7. Here we focus on one particular driver of change related to the EHR. That change (tipping point) is, we believe, the point at which the patient demands control of his or her health record. Many of us in the United States have had the time-consuming and awkward experience of requesting information from our own medical record that is often dispersed in paper form across a number of institutions. Why should we not have immediate access to our records and, if we choose, share elements of that record with whomever we see fit?

After all, many of us now spend considerable time each day in front of a computer, where we have access to our bank records and other personal data. Why should we not have that access to our EHRs and even add to our patient records ourselves to update our care providers? Consider a resource like Patients Like Me, where people choose to share and discuss their conditions with each other.[10] As more of the Web 2.0 generation (i.e., those familiar with sharing and communicating online) take interest in their EHR, the demand will likely increase for availability and access.

Before, we focused on the broader adoption of the EHR from the patient's perspective, which we see as a driver. However, savings from error reduction, apparent efficiency gains, and government regulation will all drive broader adoption of the record from the institutional perspective as well.

2.4.2 Smart Devices

The term "smart device" is catchy, but what does it mean to healthcare and the provider? If patients and care providers alike were asked, each would likely come up with different devices as examples of smart devices and of what each one means to healthcare. Let us offer a patient-centric view that suggests that it is a device trusted in some way to improve efficiency and quality of life—in some respects, an automated teller machine (ATM) for healthcare. We all trust an ATM to give us the right amount of money and update our accounts correctly. Given the ATM analogy, it is clear that such devices do not have to be "whiz-bang" new technologies; they could be something as ubiquitous as the mobile phone. In parts of the developing world, the mobile phone is emerging as a valuable tool in reporting and receiving healthcare information, so why do we not see more of this in the developed world?

With these definitions of simplicity and efficiency, a smart device might simply be a device for measuring glucose levels in an unobtrusive way. To a physician, it might be a device for good voice-to-text translation that negates the need for a transcription service. On the other hand, it could be something more comprehensive that provides the ability to access x-rays, MRIs, the latest laboratory results, or the patient history on a smart hand-held device that allows a physician's notes to be hand written and correctly understood!

All these devices are in place or on the near horizon. Once we reach the tipping point, they will become mainstream.

2.4.3 Visualization Devices

As stated earlier, healthcare is information rich, and that information must be visualized in a way that makes it most meaningful. Our example of the x-ray as an emergent technology that changed healthcare is one illustration of how visualization was a driver of change in healthcare. Today, microscopy, magnetic resonance imaging, endoscopy, and others are all forms of medical visualization in common use. A major breakthrough is not the visualization devices and techniques themselves, but rather the quality of the image, the speed of the networks, and the variety of devices from which the images can be viewed and analyzed.

Today, it is not uncommon for a tomogram to be embedded within the EHR and recalled by the patient at home or by a variety of specialists using a variety of devices, including those that are handheld. There is no technical impediment to providing this kind of visualization today—only the right price point, awareness, and desire. The future will likely include three-dimensional viewing and new forms of interaction with the images, including tactile control.

Just think what can be done today with an iPhone: Imagine that the image on the phone is a high-definition x-ray or MRI. Last, do not forget humble video. In the time that it has taken to read this section, about 16 hours of video were uploaded to YouTube. This is startling and speaks to the increasingly ubiquitous nature of video and podcasts in our daily lives. The impact of this virtual world is discussed in detail in Chapter 17; for now, consider how it can affect patient care. In the future, it may be that interactions between the physician and patient or the pharmacist and patient will be routinely captured on video and become part of the EHR for instant recall and referral.

2.4.4 Telemedicine and Telepharmacy

Telemedicine and telepharmacy are defined here as combining visualization, as discussed previously, with the use of the telephone, Internet, or other medium to provide healthcare at a distance. In the most advanced cases, telemedicine might imply surgeons performing a complex operation close to a battlefield from a site thousands of miles away by steering robotic arms to perform the procedure. A simpler and likely more ubiquitous form of telepharmacy might be a pharmacist discussing a patient's prescription on the telephone while they share data on their respective computer screens about the drug being described. Although the latter scenario is doable with the technology most of us have today, the appropriate government legislation, a business model, and the will to do it are lacking.

Telemedicine and telepharmacy are most important when a geographic barrier exists. An example is the North Dakota State University (NDSU) telepharmacy project[11]—a project in a state with a large rural population, many of whom do not have ready access to a pharmacy. To quote the university's Web site:

> A licensed pharmacist at a central pharmacy site supervises a registered pharmacy technician at a remote telepharmacy site through the use of video conferencing technology. The technician prepares the prescription drug for dispensing by the pharmacist. The pharmacist communicates face-to-face in real time with the technician and the patient through audio and video computer links. The North Dakota Telepharmacy Project is a collaboration of the NDSU College of Pharmacy, Nursing, and Allied Sciences, the North Dakota Board of Pharmacy, and the North Dakota Pharmacists Association. North Dakota was the first state to pass administrative rules allowing retail pharmacies to operate in certain remote areas without requiring a pharmacist to be present.

The preceding extract is an example of the will and the legislation being in place. With the growing use by the population at large of online, real-time communication services for voice and video (e.g., Skype), patients' demands for such services from healthcare providers can only increase in years to come.

2.5 CHANGES EXPECTED TO RESULT FROM BIOTECHNOLOGY

Traditionally, patient care as provided by the physician and pharmacist has been distinct from the research and development of products used by these care providers.

This distinction began during training and would lead to the awarding of an M.D. or a Pharm.D. degree, rather than a Ph.D. degree. Cross-training of students to receive both M.D. and Ph.D. degrees or Pharm.D. and Ph.D. degrees is an enabler of change and reflects the growing convergence of what were two distinct disciplines. Health sciences campuses around the world are introducing changes to their curricula to accommodate the emerging cross-disciplinary field of translational medicine. This field, which integrates work at the laboratory bench with the care of the patient at the bedside—or, stated more formally, the study of genotype to phenotype—is affecting and will continue to affect healthcare.

How does translational medicine affect pharmacy practice, and what is its relationship to pharmacy informatics? We will try to illustrate how these emergent disciplines and the technology associated with them are beginning to have an impact on pharmacy practice and will likely do so even more in the future. The connection to informatics comes from the large amounts of data generated by these new genotype and phenotype technologies, which only the computer can summarize for us. This new way of thinking about healthcare is being called "digitally enabled genomic medicine."

2.5.1 Genomic Medicine

The story of genomic medicine can be traced back at least to Oswald Avery who, in 1944, showed that DNA was indeed the means by which genetic traits are transferred. The double helix discovered by Watson and Crick in 1953 provided us not only with science's most well known logo, but also with insights into the structure–function relationships, and hence mechanisms, that underlie heredity and development. The culmination occurred in 2000 with the release of the first draft of the human genome (the blueprint of life)—the biological equivalent of the first moon landing.

More accurately describing the genes present in the human genome and the subsequent watershed of understanding that has arisen from the study of the human genome are starting to have and will have an ever increasing impact on illness and healthcare. With improvements in technology for DNA sequencing, the estimated $0.1 billion to $1 billion price tag for sequencing the first genome is down to $10,000 and is estimated to fall to $50 per genome in the next few years. Your complete genome sequence will likely become part of your medical record in the future. Of course, legal and ethical implications of using genomic information are being questioned and dealt with more slowly than the technology that is raising the questions.

Most popular attention is focused on the human genome; however, to scientists, the genomes of humans and many other species represent a foundation from which new understanding of the more complex features of life begins—features dubbed with different "-omics" names. For example, the genome defines our protein complement and new enabling technologies have been developed to study proteins in the field of proteomics. Proteins, DNA, RNA, and many molecules comprise a living system and it is the interaction of these components that is important. Such interactions comprise a variety of pathways for regulation, metabolism ("metabolomics"), and signaling.

By analogy, if the pathways are the wiring of the cell, then how the current flows through that wiring defines how that cell will perform. Understanding the dynamics of the living system in this way comprises the field of science called "systems biology." The ultimate goal is to simulate accurately, by computer, a living system in such a way that perturbations can be predicted and treated before serious illness arises. We are a long way from this level of understanding, but some early developments are already affecting healthcare and are discussed in the following sections.

The popular focus is on the knowledge gained from determining the sequence of the human genome; however, it is important to remember that the genomes of many other organisms have been determined or are being determined. These developments define the field of comparative genomics, which has many implications for healthcare in the future. Consider one generic approach: By knowing the genomes of a variety of pathogens (viral, bacterial, fungal) that affect human health (e.g., tuberculosis, malaria), through comparative genomics (comparing pathogen to human), we can begin to better understand the unique characteristics of the pathogen. This in turn provides opportunities to develop drugs and other treatments that specifically target the pathogen, but not the human.

2.5.2 New Modes of Diagnosis

One utility of genomic medicine is in biomarkers for the early detection of disease. Biomarkers are not a new concept. Blood pressure reading is a biomarker for possible hypertension and body temperature is a biomarker for possible fever. Prostate-specific antigen (PSA) is a protein produced by cells of the prostate gland and a well-established biomarker for abnormal prostate activity possibly indicative of prostate cancer. These examples of biomarkers represent a movement in diagnosis from phenotype back toward genotype—that is, from the complete living system back to the specific protein.

Genomic biomarkers take us further back still to the genome itself by identifying genes known to be associated with specific disease states. News articles of gene–disease associations appear regularly, but the identification of one (of possibly many) genes associated with a disease is a long way from having a practical, inexpensive test as a diagnostic tool.[12] Nevertheless, a staggering number of possible genetic tests are emerging. Gene Tests provides an up-to-date list of the genetic tests that are currently available.[13]

2.5.3 New Modes of Delivery

Here we use the term "delivery" broadly, to speak not only of drug delivery, but also of delivery of any kind of healthcare. However, let us start with drug delivery. A pharmacy student is taught early on that the effectiveness of a potential drug involves more than how well it binds to its receptor. There are issues of absorption, distribution, metabolism, and excretion (ADME). It therefore makes sense to try to have the drug reach the site of action without adversely encountering ADME issues. Nanosized devices capable of moving through the bloodstream equipped with implanted controlled-release mechanisms, perhaps through radio control, are examples of controlled-delivery devices to better reach the site of action.

Nanoparticles are microdevices at one end of the size spectrum; at the other end are the macrodevices, such as monitoring devices, which also deliver better healthcare. Although monitoring of vital signs is routine, more extensive monitoring devices that better monitor blood sugar levels or even hormone and metabolite levels are likely to become commonplace. In the future, we will begin to see monitoring devices that track progression or regression of disease reaction to therapeutic intervention in real time. Again, this provides a mass of new information that will figure into the life of the pharmacist in the coming years.

2.5.4 New Modes of Drug Discovery

Drug discovery is a broad and complex topic. The purpose here is simply to stimulate thinking about the changes that are likely through new biotechnologies, the impact they will have on pharmacy practice, and the ever increasing need for pharmacy informatics in the drug discovery process. The traditional idea in drug treatment is to find one drug that binds to one receptor and treats one disease. As the complexity of the living system is slowly revealed, this viewpoint is proving naïve. We are treating a living system that has evolved over at least 3 billion years. Thus, it is not surprising that very few foreign substances are found to be therapeutic. Rather, the living system has evolved defense mechanisms to protect itself against such substances. In a pragmatic way, this is reflected in the "rule of five" that defines what constitutes a likely pharmaceutical.[14]

Because the living system has evolved to be synergistic with the environment, it is not surprising that natural products often prove to be successful therapeutic drugs. In the period from 1981 to 2006, 974 small-molecule new chemical entities were introduced; 63% were naturally derived or semisynthetic derivatives of natural products. For certain therapeutic areas, such as antimicrobials, antineoplastics, antihypertensives, and anti-inflammatory drugs, the percentages were even higher. Despite the implied potential, only a fraction of Earth's living species has been tested for bioactivity. This situation will likely change in the coming years as a result of metagenomics[15]—a field of science that performs multispecies genomic sequencing directly from environmental samples.

The elucidation of the human genome now provides us, in principle, with the "drug-gable" genome—all the likely drug ligands and receptors. We say "in principle" because many of the protein coding regions within the genome have yet to be annotated and hence identified as likely receptors. Again, there is a certain naiveté in this thinking. Who is to say the drug binds to only one receptor? The idea of polypharmacology (polyvalent/covalent), in which a given drug binds to multiple receptors to lead to a collective multivalent outcome, seems more appropriate.

A broader than expected affinity by a drug such that it binds to multiple receptors can be both a blessing and a curse. It is a blessing because it may provide multiple points to effect a positive outcome on the patient—the notion behind so-called dirty drugs. It is also potentially a curse because it may result in adverse unanticipated side effects that are not revealed until late in the drug development process. Torcetrapib, a cholesteryl ester transfer protein inhibitor to reduce serum cholesterol that was developed over a period of 15 years at a cost of $850 million, is a case in point. Stage III clinical trials revealed that the drug caused fatalities attributed to hypertension—an unanticipated side effect attributed to off-target binding to a number of receptors other than the single intended receptor.

2.5.5 Personalized Medicine

The realization that patients respond differently to the same dose of the same medication has been known for a long time. In 1902, Archibald Garrod first asserted the hypothesis that genetic variations could cause adverse biological reactions when chemical substances were ingested.[16] He also suggested that enzymes were responsible for detoxifying foreign substances, and that some people do not have the ability to eliminate certain foreign substances from the body because they lack enzymes required to metabolize these materials.

Drug reactions based on inherited traits were first recorded during World War II, when some soldiers developed anemia after receiving doses of the antimalarial drug primaquine. Later studies confirmed that the anemia was caused by a genetic deficiency of the glucose-6-phosphate dehydrogenase enzyme. Similar reactions to succinylcholine and isoniazid were studied and revealed that deficiencies in enzymes led to an inability to metabolize these drugs normally. After studying adverse drug reactions to primaquine, succinylcholine, and isoniazid, Arno Moltulsky proposed in 1957 that inherited traits may not only lead to adverse drug reactions, but may also affect whether the drugs actually work.

Today, the study of this varying genetic disposition to different pharmaceuticals is called pharmacogenomics or pharmacogenetics. The growing body of information on this field is maintained in a database called PharmGKB (the Pharmacogenomics Knowledge Base).[17] The database can be searched in various ways—for example, according to different levels of biological complexity: gene, protein, pathway, drug. Thus, the search can be conducted by known pharmacogenomics associated with a specific drug, the genes involved, the pathways that contain those genes, the literature associated with the biology, and clinical trials offered as evidence for the genetic disposition.

Pharmacogenomics represents an added stress on the pharmaceutical industry because it is more profitable to sell one drug to a larger population than to have a variety of drugs and doses for subsets of the population. Notwithstanding, personalized drug treatment is a reality that affects pharmacy practice. Consider a recent illustration. The International Warfarin Pharmacogenetics Consortium and members of PharmGKB introduced a warfarin dosing regimen based on both clinical and genetic factors.[18] The FDA has changed warfarin's package insert to reflect this new pharmacogenomic knowledge.

Personalized drug treatment is part of a broader field of personalized medicine that moves us away from medical practice that is based on overall standards of care defined across large cohorts of patients. Tracking and responding appropriately to care that is defined for individuals rather than cohorts require a new level of information processing; as such, it is a driver of change that again highlights the growing importance of pharmacy informatics.

2.6 A FINAL REALITY CHECK

After reading this brief introduction to the many changes in information technology and biotechnology that are underway, it would be easy to imagine that pharmacy practice will be very different in the future. Chapter 18, which concludes this book, is a reality check on the current situation in pharmacy practice and what we can expect. Taken together, the conclusion is that change will come, as driven by some of the developments described in this chapter; however, change will be slower than defined by technology alone because of the need to overcome legislative, economic, and sociological barriers that slow the process markedly.

In other words, for pharmacists this represents a journey that has begun and will be marked by events such as broader use of the EHR and the complete annotation of the human genome. These and other events are information driven, so the need for pharmacy informatics can only increase. That increase will be associated with a set of core values that lead to improved patient management, outcomes, and overall improvement of health. Examples of improvements that can be expected to affect pharmacists include:

- reduction in prescribing errors;

- prevention of adverse drug–drug interactions;

- improvements in communications between patients and pharmacists by removing geographic barriers; and

- treatment regimens based on genetic disposition.

These are exciting and important changes in which the pharmacist can contribute to improved healthcare.

REFERENCES

1. Gladwell, M. *The tipping point: How little things can make a big difference.* Boston: Back Bay Books, 2002.
2. Hillestad, R., and Girosi, F. Advanced care: The promise of health information technology for cost, quality, and privacy. The Rand Corporation. 2009 (www.rand.org/publications/randreview/issues/spring2009/hit.html).
3. Centers for Medicare and Medicaid Services (www.cms.hhs.gov).
4. Commission on a High-Performance Health System. The Commonwealth Fund. (www.commonwealthfund.org/Content/Program-Areas/High-Performance-Health-System/Commission-on-a-High-Performance-Health-System.aspx).
5. World Health Organization. Health statistics and health information systems (http://www.who.int/healthinfo/statistics/en/).
6. World Health Organization. The world health report. United States of America (http://www.who.int/countries/usa/en/).
7. World Health Organization. The world health report (www.who.int/whr/2000/media_centre/press_release/en/index.html).
8. Tufts Center for the Study of Drug Development (csdd.tufts.edu/InfoServices/OutlookPDFs/Outlook2009.pdf).
9. Wikipedia. VistA (en.wikipedia.org/wiki/VistA).
10. Patients like me (www.patientslikeme.com).
11. Telepharmacy. North Dakota State University (www.ndsu.edu/telepharmacy).
12. Pritzker, K. P., and Azad, A. Genomic biomarkers for cancer assessment: Implementation challenges for laboratory practice. *Clinical Biochemistry* 2004. 37:642–646.
13. Gene tests. University of Washington, Seattle (www.genetests.org).

14. Wikipedia. Lipinski's rule of five (en.wikipedia.org/wiki/Lipinski%27s_Rule_of_Five).
15. Wikipedia. Metagenomics (en.wikipedia.org/wiki/Metagenomics).
16. Mancinelli, L., Cronin, M., and Sadee, W. Pharmacogenomics: The promise of personalized medicine. *AAPS PharmScience* 2000. 2:E4.
17. Klein, T. E., Chang, J. T., Cho, M. K., et al. Integrating genotype and phenotype information: An overview of the PharmGKB project. Pharmacogenetics Research Network and Knowledge Base. *Pharmacogenomics Journal* 2001. 1:167–170.
18. Klein, T. E., Altman, R. B., et al. International Warfarin Pharmacogenetics Consortium. Estimation of the warfarin dose with clinical and pharmacogenetic data. *New England Journal of Medicine* 2009. 360:753–764.

II

Prerequisites

CHAPTER 3

Computer and Informatics Basics

Philip E. Bourne

CONTENTS

3.1 INTRODUCTION

Whether we use a computer in pharmacy practice or are involved in the nuts and bolts of developing or maintaining a pharmacy information system, computer and information management basics affect us all. It is this fundamental information that is the subject of this chapter. This topic is not easy to write about for at least two reasons. First, information technology moves very rapidly. Consequently, any specific details provided today will be out of date very soon. Second, students of pharmacy become better prepared each year to handle information technology as it increasingly becomes a regular part of their lives. This second point takes us to the first lesson.

3.2 LESSON ONE: HAVE COMPASSION FOR THOSE LESS KNOWLEDGEABLE ABOUT THE TECHNOLOGY

Let me illustrate the importance of this with an example. Many years ago I ran a computer facility where a group of basic and clinical science researchers shared a computer that was less powerful than a laptop is today. One evening I received a call from one of the scientists, who said, "I think the computer is down." I hit the enter key on my keyboard connected to the same computer and, yes, it appeared down. I responded, "Yes it appears down." There was a pause and then the scientist said, "Is the information still in the wires?" I quipped back, "Let me go and take a look." This would perhaps be laughable if not for the fact that the scientist in question went on to win a Nobel Prize a few years later.

In summary, very smart people can often know little about information technology. Because some of them will be customers and patients, they deserve to be respected and guided as needed to find the information they need to make informed decisions about their healthcare. This is a recurring theme of this book.

Now that we have learned to approach the topic with compassion, let us first look at some of the trends.

3.2 LESSON TWO: TECHNOLOGY WILL ADVANCE AT A SPEED AND IN WAYS THAT CANNOT BE PREDICTED

We are all very poor at predicting the future. Consider Tim Berners-Lee,* the inventor of the World Wide Web, who had no idea of what would become of his innovation. No one would have predicted that pharmacy practice, like so much of our personal and professional lives, now uses the Web. Likewise, Thomas Watson, president of IBM, said in 1943, "I think there is a world market for maybe five computers." All of this is nicely summed up in *The Next Fifty Years. Science in the First Half of the Twenty-First Century,*[1] where a number of major contributors to science and technology do a great job of predicting what might be, but rarely venture beyond 5 years and certainly not 50.

I said that we cannot reliably predict the future of information technology; however, success in pharmacy informatics may require doing so in ways that range from the

---

* Most pharmacy students have no idea who invented the Web, which speaks to the ubiquitous nature of how technology is perceived. There is a brief "wow" moment, followed by immediate acceptance and a hankering after the next new innovation.

purchase of a new laptop computer to making decisions about the hardware and software needed to support a major pharmacy chain. How should this apparent dilemma be approached?

A useful way would be to examine data that represent trends from the past that can be projected into the future. Perhaps the most well known trend in information technology is attributed to Gordon Moore and is known as Moore's law (http://en.wikipedia.org/wiki/Moores_law), which states, "Since the invention of the integrated circuit in 1958, the number of transistors that can be placed inexpensively on an integrated circuit has increased exponentially, doubling approximately every two years." Simply stated, if a new computer is bought every 2 years, the new one can be expected to be twice as powerful as the one bought previously.

This trend is expected to continue. It is hard to predict for how long, but this is perhaps one of the major features that defines how information technology does and will have an impact on the healthcare industry. Similar trends can be seen in disk storage, display technology, network speed, etc. Accompanying speed, storage capacity, and screen quality are smaller footprints that can change our habits, moving us away from desktops to increasingly sophisticated handheld devices, with names like personal digital assistants (PDAs) and smart phones as exemplified by Apple's iPhone (see Chapter 13).

3.2.1 A Few Useful Metrics

Given the speed of change, it makes no sense to quote absolute numbers (they will be outdated before this book is published). Rather, it is a better idea to point to metrics that are important to consider when choosing information technology to support needs in pharmacy practice.

3.2.1.1 Processors

The clock rate is the most important metric in defining the speed of a computer. In simple terms, it is the time it takes to flip a zero into a one and vice versa—the basic operating instruction of a computer. Thus, a clock rate of 3.0 GHz means this happens three billion times a second. Choosing a computer with the highest clock rate at any given point in time will cost the most and may be overkill, depending on how the computer will be used. Talk to people and read about those performing similar tasks using computers of different clock rates. A faster rate may be a waste of money. As when you shop for a car, do not be lured by the fastest, but rather by that which gets you there the most economically and reliably.

3.2.1.2 Physical Memory

Physical memory is frequently more important than processor speed in defining how well a computer will function on the applications that are used regularly. Today, memory is typically in the 1–10 gigabyte (GB) range, with 1 GB meaning one billion bytes can be stored. A byte is eight bits (ones or zeros) and 1 GB is equivalent to one billion bytes. Too little memory does not mean that applications will not run, but rather that they will run more slowly as a result of paging and swapping, which are processes controlled by the computer's operating system (e.g., Windows, Linux, or Mac). They share the memory among the applications running by copying parts of memory to the computer's hard disks and then copying them back—an inefficient process. The hardware requirements are usually specified for the software purchased (or downloaded) so be sure to have enough memory for the applications to be run.

3.2.1.3 Graphics Cards

Also known as the video card, this is an additional hardware component responsible for a number of applications involving display and rendering (manipulation) of graphics and video. These days, almost any graphics card will perform well for typical applications; however, for gaming and, more importantly, applications for manipulating digital x-rays, MRI scans, etc., the type of card may be important. In most cases, the software used will dictate the best graphics card to purchase.

3.2.1.4 Disk Storage

It is perhaps the revolution in disk storage more than any other hardware component that has driven the IT revolution and is having an impact on pharmacy practice. As I write this, Gmail, Google's free e-mail offering, is registering that over 7,000 terabytes (TB) of disk storage are available for users worldwide to store their e-mail freely. At the same time, YouTube is receiving 14 hours of video uploads every minute, all stored free.

How can this be possible? In part this is because Moore's law holds for disk storage, too, and partly because new business models have emerged that use these cheap resources to deliver services for the opportunity to advertise to the user. For those of us who remember the days of floppy disks and 1 MB storage, this is indeed a revolution—no pun intended.* Laptops now typically have disk storage of over 100 GB and the main issue becomes maintaining the integrity of the data (which we will get to subsequently). When disk storage is considered, the following issues become important in addition to capacity:

- rotational speed—the faster a disk rotates, the faster data can be read or written;

- average seek time (in milliseconds)—the time it takes to find data on the disk;

- temporary buffer size—a cache of memory to maintain data recently read from or written to the disk;

- mean time between failures (MTBF)—the reliability of the disk, because it is a mechanical device, after all. Solid-state storage (similar to physical memory and often called chip-based hard drives) is more expensive but has no moving parts and will have fewer failures†;

---

* Disks are currently mostly mechanical with revolving parts (hence the pun and a recognition that mechanical items fail).
† Compare the original iPod to the iPod Nano.

- hot swapping—more important in desktops and servers in order to be able to swap a disk out without powering off the computer or rebooting; and

- random array of inexpensive disks (RAID)—serves the purpose of generating a large amount of redundant storage for large applications like maintaining a pharmacy information system.

3.2.1.5 Network Cards

Network cards are of two types: hardwire and wireless. Laptops typically contain both (do not buy a laptop without a built-in wireless card); desktops and servers typically have only hardware, but wireless can be added using a USB port (see next section). Speed of the card is rarely an issue if a recent card is used because the bottleneck to network speed invariably lies elsewhere. Each network card has a unique physical Ethernet address, which ultimately is what identifies an individual computer among the many millions accessing the Internet at any given moment. The address is represented as six parts, each consisting of two hexadecimal numbers (e.g., 08:00:20:03:72:DC). We will come back to this shortly.

3.2.1.6 USB Ports

We are in a fortunate era in computing where most hardware manufacturers support universal serial bus (USB) adaptors that make it simple to add a variety of peripheral devices to any computer. The number of USB ports on a computer can be an important factor when purchasing a desktop or laptop, as can the speed of these ports. To date, there have been three USB releases, with each one markedly faster than the previous. Be sure that the computer used or to be purchased has the latest release.

3.2.1.7 Firewire Ports

Firewire (also called IEEE 1394 interface, i.LINK, and Lynx) is a frequently used video specification typically found on camcorders and other devices. It is not necessary to have a firewire port because firewire to USB cables is very common and typically shipped with any purchased video equipment.

3.2.1.8 DVDs/CDs

DVDs and CD read/write devices are standard on many laptop and desktop computers. CDs store up to 700 MB; DVDs can store between 4 and 17 GB and are more useful secondary backup storage devices for materials that are accessed occasionally. Faster media such as USB drives are better for materials accessed more frequently. Storing movies, photos, and other media files on DVDs is typical. Speed of DVD writing is expressed as a number times the speed of the original drives from around 1995. Drives at the time of writing will typically exhibit write speeds of 20× their original counterparts.

3.3 LESSON THREE: COMPUTERS FAIL, SO BE PREPARED

Students often come to me in disbelief that their computers have failed. There seems to be a blind and misplaced belief that technology is invincible. Yes, the MTBF of any computer component has been increasing over the years, but failures still do occur. The most common and most devastating failure is that of the disk drive. This is not surprising because it is the mechanical part of most computers.

3.3.1 Backup Strategies

This is nothing more than common sense. Ask what the implications are if the disk fails now. If the answer is, "I will lose important data forever" (photos, etc.), an appropriate backup strategy is not in place. If the answer is, "I will only lose what I did in the last few hours," then the personal strategy is probably sound. Strategies for businesses, including running pharmacy systems, obviously need to be more stringent because their survival could well depend not only on having good backups of, for example, patient data and transactions, but also on the ability to restore the exact state of what has transpired very quickly when a failure occurs. Let us consider personal and business strategies briefly and separately.

3.3.1.1 Personal Backups

As I write this chapter, I have Microsoft Word set to autosave my work every 10 minutes (the default). Thus, if the computer crashes as a result of a fault other than a disk failure, I have only lost a maximum of 10 minutes' work. Typically, I save my work to an alternative media every hour or two to safeguard (at least what I regard as) my intellectual output. This can be done in various ways and computing trends in general are making this easier than ever with a move toward central data servers. Here are four possibilities:

1. Save the same information to a different physical device attached to the same computer. This could be a USB (aka thumb) drive or a larger external USB drive. Additionally, data can be saved to a CD or DVD. Physical proximity still represents a potential point of failure in the case of more unlikely problems like fire and flood, but they happen.

2. Save the information to a central server in the workplace if one is available. This is preferable (and, in fact, may be the only acceptable option) to the following options if the data need to be secured.

3. E-mail materials to yourself where you have an e-mail account on a remote server (e.g., Gmail or Yahoo mail). This is easy for a few relatively small files, but not suitable for a large number of large files.

4. Use Google Docs or another central free file storage service. Although they are intended as sites for shared documents, they work for backups, but again more for a small number of relatively small files.

For me, a combination of all four options works. For large collections (e.g., photos and music), I use the first option; for important documents, I use a combination of the third and fourth options. Using a compression program like gzip or WinZip allows compression of a large number of files into a single file of reduced size. This helps in keeping track of them and facilitates sending them over the Internet.

3.3.1.2 Business Backup Strategies

There are many possibilities for backing up business files. For example, consider the RSCB Protein Data Bank (PDB; www.pdb.org), which is a 24/7 service where the integrity of data is critical. The PDB is an important resource for scientists engaged in drug discovery and other related tasks in molecular biology.

The PDB comprises about 200 GB of data and software; the primary access site (www.pdb.org) is composed of five independent servers, each with a complete copy of the PDB. Redundant load balancers (if one fails, the other takes over) direct incoming Web traffic to the most underused of the servers in the pool. This configuration of three servers is replicated at two other physically (and, in one case, geographically) remote sites. An independent service, called UltraDNS, pings (communicates with) the three primary servers every few seconds. If they do not respond, the address www.pdb.org is redirected to one of the other locations; if that one does not respond, it will go to the third.

In this way, because the systems are identical, the user is unaware that anything has happened. Clearly, this provides data integrity through redundancy, but also illustrates how a 24/7/365 system uptime can be maintained. Over the past 10 years of PDB operation, the system has had greater than 99% uptime and never lost a piece of data or software. Hospitals, pharmacies, etc. have variations on this type of scheme to maximize uptime and data integrity.

3.4 NETWORK BASICS

We have already introduced the concept of a physical address as a unique identifier fused into the network card of a computer. Let us now consider a few other terms and see how they all work together to make the Internet the amazing resource that it is (see Figure 3.1):

- *Physical address* (also known as a media access control or MAC address) is unique to a computer forever (unless the network card is changed). This is a six-part hexadecimal number (e.g., 00-12-3F-52-B5-29).

- *IP address*—either static or dynamic—is how the Internet recognizes a computer. This is a number of the form w.x.y.z (e.g., 192.168.1.100).

- *Hostname* is a name that a human can use to identify the computer.

- *Domain name server (DNS)* relates the hostname to the IP address (e.g., 68.105.28.11); usually, there is more than one, in case one fails. Humans remember names much better than numbers. Computers like numbers much better than names. DNS translates (maps) names to numbers (IP addresses) like Google.com to 74.125.45.100; this is a system much like a phone book. Important Web sites (like Google and the PDB) have more than one IP address in case one fails.

- *Subnet mask* limits the search space to resolve IP addresses. IP addresses are really two addresses combined into one. These two addresses are a network address and a host address. The network address defines a group of computers (like the PDB) and the host defines a computer within the group. This is much like the mail I receive at

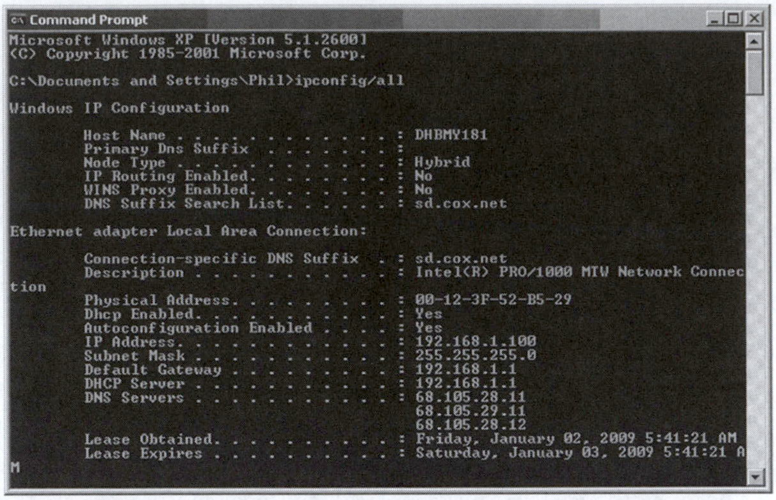

FIGURE 3.1   Details of an Internet connection as shown by the command ipconfig/all on a Windows XP system.

my home (the network address), some of which is for me and some for my wife or children (host address). The subnet mask is used to separate an IP address into its network address and host address components. An example of a subnet mask is 255.255.255.0.

In order to communicate over the Internet, a computer must have an IP address linked to the MAC address fused into a network card. There are two types of IP addresses: static and dynamic. Static IP addresses are assigned to a computer and never change; this is useful if other computers, such as a print server or a file server, always need to know where the user is. However, most computers are assigned (leased) an IP address from a pool of addresses when they start up. When a computer is turned on, the network interface card communicates with the Internet service provider (ISP) asking to borrow an IP address for a while. The ISP then selects an IP address from a pool and leases it to the requestor while the computer is working. When the requestor disconnects from the Internet or shuts down the computer, the address is returned to the ISP to be allocated to someone else.

Beyond identifying a computer on the Internet, consider how information gets transferred, such as when an e-mail message is sent. The message is sent as a series of packets of information; each can be uniquely identified with respect to where it has come from and where it is going, as well as the sequence of the packets such that the information can be reassembled in the right order. These chunks of information may not travel by the same route and certainly may not arrive at their destination in the right order, but the underlying Internet software knows how to interpret them because they use an agreed-upon protocol.

The protocol is a type of contract between sender and recipient. Different forms of transmission use different protocols. For example, Web pages are transferred using hypertext transfer protocol (http), which Web browsers recognize. Protocols such as http are layered on top of the basic transmission control protocol/Internet protocol (tcp/ip). Details of the

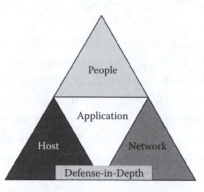

FIGURE 3.2   Components to consider for information assurance.

protocols are beyond the scope of discussion here, but a basic understanding of Internet transmission helps us understand what might go wrong. For example, security violations as a result of "packet sniffing" may occur. On a wired connection, someone can maliciously read your packets as they go by; packets sent from a laptop in a coffee shop on a wireless network can be siphoned off and interpreted by a malicious third party more easily. We will see how to safeguard your transmissions as part of what is termed *information assurance*.

3.5 INFORMATION ASSURANCE

In a nutshell, information assurance is ensuring that your information is where you want it, when you want it, in the condition that you need it, and available only to those that you want to have access to it. Clearly, this is an important issue with respect to patient privacy and confidentiality and we will revisit this issue a number of times throughout this book. Here, we are concerned with the core fundamental issues of ensuring the integrity of information as it relates to the basic components used to maintain the information.

Figure 3.2 illustrates the components that must be considered. Let us go over them one at a time, further breaking the components into smaller parts.

3.5.1 Network

We have already seen how the network can be compromised. A typical means of preventing packet sniffing is to encrypt the information. However, let us start at the level of the network itself. Institutions like hospitals and pharmacies typically have a firewall to prevent unwanted access. A firewall can be hardware, software, or both and has the job of fending off unwanted intrusions. Such intrusions can come from a user or an application program. Firewalls typically limit the users and the applications that can pass through from the external to the internal network. In this way, information can still flow freely around the internal intranet, but information coming from the Internet is very carefully scrutinized.

If the organization seeking to protect itself is small—a doctor's office, for example—it may not have the opportunity to maintain a separate intranet. In these instances, it is possible to establish a virtual private network (VPN). The VPN uses the same infrastructure as the regular Internet, but encrypts the data just on that part of the network that resides within the organization.

3.5.2 Host

The host refers to any computer on the network containing information that must be kept secure. This ranges from a personal laptop to large, central, patient database servers. Regardless of the computer, a few fundamentals are highlighted here. Although it appears obvious, physical security is often what is violated. Maintaining passwords that are difficult to guess, having secure user accounts on laptops, and not storing passwords, including those maintained by the Web browser for common applications, are good steps to take when information assurance is needed. Computers with sensitive data should not allow public access, but rather require all users to log in on their own accounts.

Likewise, vulnerabilities in the host operating system, whether it is Windows, MAC, Linux, or another operating system, are constantly being discovered. It is important to check for updates ("patches") constantly. On laptops and desktops, such checking is easily supported and should be enabled. On hosts accessed by multiple users, it is important to monitor access daily. Logs are provided for applications as well as for users accessing the system and they need to be reviewed carefully for signs of intrusion.

3.5.3 People

People are probably the weakest link in the security triangle and yet the component that gets the least attention. Students likely signed an agreement for acting so as to maintain the security of information, yet few of us remember doing so and even fewer of us could recall details of what we agreed to do. The same could well be true when working for a company where information security is important. The most obvious aspect is maintaining passwords that are difficult to guess and changing passwords frequently. Most systems and applications force this upon us in any case. Keeping passwords difficult to guess and changing them frequently make them easy to forget; however, we must resist the temptation to write them down anywhere near the computer.

3.5.4 Applications

Information assurance vulnerabilities associated with applications are most often associated with applications that are part of the computer operating system and come in various well-publicized forms. A virus is a self-reproducing application that spreads by inserting copies of itself into other applications or documents. There have been some devastating examples of viruses; one of the most well known, the "I Love You" virus, cost an estimated US$5–10 billion in damage worldwide. Beyond money value for me were some photographs that I had not backed up and thus were destroyed. Worms are a subclass of virus that can travel without help from a person and "tunnel" into a system. Last, a Trojan horse tricks the user into willingly, but unknowingly, letting a computer be infected, as were the people of Troy.

Dealing with such infiltrations is a constant battle between the perpetrator and the software manufacturer against which the attacks are aimed. New vulnerabilities are tracked by the computer emergency response team (CERT) as they are discovered. Maintainers of data centers constantly monitor CERT advisories and decide on what actions to take.

Makers of antivirus software do the same. An individual's best defense is to maintain up-to-date antivirus and antispyware software. Such software will identify these new vulnerabilities and usually prevent such intrusions from reaching a computer.

If a computer does get infected, these programs will isolate (quarantine) the infected files so that they do not do any damage. Damage can range from the need to erase the disk completely and reinstall the system software to making a computer perform slowly. Either way, this can be a serious waste of time. Operating systems define the appropriate levels of security to enforce and, depending on the importance of the information, the appropriate level of security should be enabled.

3.5.5 Encryption

With the need to protect patient privacy and hence the information about that patient maintained by the hospital or pharmacy, it is important to raise awareness and at least instill a basic understanding of encryption. The increasingly distributed nature of the healthcare system means that private information is flying around with and without wires. We have already seen how such information can be snagged by malicious intruders intercepting the information via packet sniffing. Encryption simply means, "Okay, so you have got the private information, but we are not going to let you interpret it." Encryption is almost as old as information itself; it implies a contract between the transmitter and receiver of information that allows them, and only them, to interpret the information being transmitted.

Pharmacists are likely to encounter the products and terminology of encryption such as secure socket layer (SSL) and Web addresses of the form "https://." Perhaps the most common form of encryption is public-key encryption, which uses a combination of a private key and a public key; the keys provide the means to lock and unlock the information. Only the computer knows the private key, but it gives the public key to any computer that wants to communicate securely with it. To decode an encrypted message, a computer must use the public key, provided by the originating computer, and its own private key. A very popular public-key encryption utility is called "pretty good privacy" (PGP), which allows encryption of almost anything. More information about PGP can be found at http://www.pgp.com, but the basic idea is shown step by step in Figure 3.3.

The bottom line in this approach is that a message is never sent to host B until it indicates that it will receive an encrypted message by sending a unique public key. When that public key is returned as part of the session key, only the private key for host B can decode it.

The question that remains is how to know that computer A can be trusted. Digital certificates are familiar to most computer users who download from the Internet. A digital certificate is maintained by a third party and computers A and B would need to register and be known to that third party for the message to be transferred. The need to be identified and trusted by the third party removes the possibility of communication with a bogus computer. Many of us accept trusted certificates from parties such as Sun Microsystems for Java applications and Microsoft for its applications.

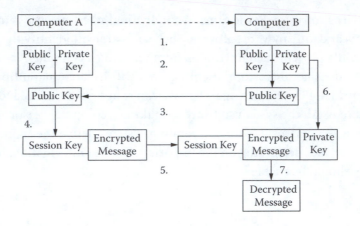

FIGURE 3.3   Encryption: (1) Computer A prepares to communicate an encrypted message to computer B. (2) Each computer has both public and private encryption keys. (3) The public key is sent to computer A, which is going to send the message. (4) The public key is converted into a session key, which contains the public key for just that one session and the encrypted message based on that public key. (5) The encrypted message and session key are sent to the computer for which it is intended. (6) Receiving computer B's private key is then used. (7) The message is decrypted.

3.6 SUMMARY

Our three basic lessons were that (1) understanding of information technology varies widely, (2) it is impossible to predict the future of information technology, and (3) bad things will happen. Here are a few basic rules for any laptop or desktop user to abide by to help keep a computer working efficiently and productively. Most of this information is just good practice whether it is applied to pharmacy informatics or not:

- Keep alert to changes in information technology that will likely have an impact the practice of pharmacy.

- Be able to restore an operating system and major applications from their original media should the need arise through disk failure or virus intrusion.

- Make regular backups of any files that cannot be lost. It helps to think, "What will I lose if the disk fails?" because, sooner or later, it will fail.

- Make sure that the software firewall is turned on.

- Make sure that the latest operating system patches have been obtained, unless there is a compelling reason not to allow for automatic updates.

- Purchase antivirus software and run it at least once per week. This will protect against new viruses as they are released.

- Install antispyware software, particularly if data exist that others should not read.

- Have a password set on the computer to prevent physical access by others.

- Keep browser security settings high (for Internet Explorer, select default level) or, at a minimum, medium.

- Never open mail attachments unless the sender and content are known.

- Do not click on pop-up ads. They can download software (e.g., Trojan horses) to a computer.

- Do not download any applications that cannot be trusted.

REFERENCE

1. Brockman, J. E. *The next fifty years. Science in the first half of the twenty-first century.* New York: Vintage Books, 2002.

CHAPTER 4

Controlled Vocabularies

Philip E. Bourne and Susan M. McGuinness

CONTENTS

4.1 INTRODUCTION

Computers are not as smart as humans—not yet anyway. Although the IBM Deep Blue computer was able to beat grand master and former world champion Gary Kasparov at chess in 1997, it was accomplished by computing possible moves rather than applying what we would call human interpretation and intuition. By the time this book is read, it may well be that IBM's next challenge, to win a game of *Jeopardy,* will also have been met, taking us yet closer to the day when computers think like humans.

Humans make sense of things through the use of controlled vocabularies and standards. A controlled vocabulary is a collection of words with specific meaning; a standard is a collection of rules that determine how the words in the controlled vocabulary can be combined and used for a purpose. The English language can be thought of as one standard, with the controlled vocabulary being the words that comprise the language. Thus, the most obvious instance of a controlled vocabulary can be found in a dictionary. For example, the *Oxford English Dictionary* defines the words by which we communicate and provides the basis by which we understand each other. It is no surprise, then, that if computers are to communicate and to reason, they too must use standards and controlled vocabularies. Because computers do not have the benefit of five senses, it becomes all the more important to make standards and vocabularies as comprehensive and machine usable as possible.

Consider the statement "27." Without context, even a human would be unable to comprehend what it means. However, "the temperature is 27" provides enough information for us to understand what is being said, but not for a computer. We could look around, see we were standing in Alaska, and could infer that the statement meant "27 degrees Fahrenheit" from the context (i.e., Alaska). A computer could not do the same. If this fact were entered into a computer, the computer would need to be told that the data value 27 is associated with the data name, "temperature," and that the units of temperature are "Fahrenheit."

In one sense, these are metadata–data about data. Metadata are critical for most forms of data processing using the computer. In this simple example, the standard is the Fahrenheit measure of temperature and the controlled vocabulary term is how this data value is named. "Temp.," "ambient temperature," "temperature," "hot," and "cold" are all ways for a human to indicate the same thing; however, for a computer to identify the similarity, a single term must be used consistently to represent the concept of temperature. That single term is part of a controlled vocabulary or, simply, a vocabulary. Much of the information in later chapters concerning pharmacy and hospital information systems involves adopting and enforcing standards and controlled vocabularies. Sometimes the terms "standard" and "vocabulary" are used interchangeably, but in principle, multiple vocabularies may conform to a given standard. What are important to grasp are the concepts of the need for standards and vocabularies, what some of them are, and how they are used.

4.2 STANDARDS

To be more specific, a standard can be considered:

- something that facilitates a coordinated action;
- something that fosters uniformity;
- something critical for the exchange and effective use of information;
- a means by which standard terms (vocabularies) can be expressed; or
- in pharmacy practice, something that fosters the organized treatment of patients.

Examples of standards and the organizations responsible for these standards used elsewhere in this book include:

- ASC X12, the Accredited Standards Committee (www.x12.org), is an organization that develops electronic data interchange (EDI) standards and related documents for national and global markets, including pharmacy.

- ASTM–ASTM International (www.astm.org) provides technical standards for materials, products, systems, and services.

- HL7, Health Level 7 (www.hl7.org), provides standards for interoperability that improve care delivery, optimize work flow, reduce ambiguity, and enhance knowledge transfer among all stakeholders, including healthcare providers, government agencies, the vendor community, fellow standards-developing organizations (SDOs), and patients.

- NCPDP, National Council for Prescription Drug Programs (www.ncpdp.org), is an organization that develops standards for all areas of the pharmacy services industry.

- NGC, National Guideline Clearinghouse (www.guideline.gov), is a public resource for evidence-based clinical practice guidelines.

4.3 VOCABULARIES

Controlled vocabularies are defined to conform to a standard. Unlike standards, which are static once they are defined, the vocabularies conforming to that standard need to be dynamic to embrace new terminology that enters the field in the same way that new words enter the English language. In principle, terms never leave the vocabulary. In this way, the use of a term can always be tied to a definition or point of reference. Using the English language analogy, a word may go out of common usage and, in principle, could be removed from the dictionary. However, when someone reading an old book written when the term was in common use may need to look up its meaning somewhere, and hence the term should be preserved in the dictionary. Examples of vocabularies used in pharmacy practice include:

- CPT (current procedural terminology; www.ama-assn.org/ama/no-index/physician-resources/3112.shtml) is the most widely accepted medical nomenclature used to report medical procedures and services under public and private health insurance programs. CPT is maintained by the American Medical Association.

- GO (gene ontology) is used to describe gene products (www.geneontology.org).

- ICD (international classification of diseases; www.who.int/classifications/icd/en) classifies diseases and other health problems recorded on many types of health and vital records, including death certificates and health records. It is maintained by the World Health Organization.

- LOINC (logical observation identifiers and codes; www.loinc.org) provides universal codes and names to identify laboratory and other clinical observations.

- NDC (national drug code; www.fda.gov/cder/ndc) is a unique, three-segment number used as a universal product identifier for human drugs and supported by the U.S. Food and Drug Administration (FDA).

- RxNorm (www.nlm.nih.gov/research/umls/rxnorm) provides normalized names for clinically used drugs and links its names to many of the drug vocabularies commonly used in pharmacy management and drug interaction software.

- SNOMED (systematized nomenclature of medicine—clinical terms; http://www.nlm.nih.gov/research/umls/Snomed/snomed_main.html) is the most comprehensive vocabulary of clinical terms.

Vocabularies become particularly important when information is added to and subsequently retrieved from a database. Before describing the use of controlled vocabularies in databases, we must define databases.

4.4 DATABASE FUNDAMENTALS

4.4.1 Database Structure

A database is an organized collection of information on a specific subject. Telephone books, library catalogs, and hospital and pharmacy information systems (see Chapters 6 and 7) are all examples of databases. The subject of a database could be scholarly literature, protein structures used in drug discovery, or a set of patients. Most databases described in this textbook are relational databases, which consist of a set of related tables that look like individual sheets in a spreadsheet. Each row in a table (also called a record or tuple) represents one instance of the entity described by the table. A hypothetical inpatient hospital patient database might contain a table of patient information, a table of drug information, and a table of physician information. These tables would be related because physicians treat patients and prescribe drugs, and patients have physicians and are taking drugs.

Each record in the patient table has information about one patient. The columns of a table are called fields. Each record consists of a collection of fields that include data describing the attributes of that record. Fields in our patient table might include name, sex, age, height, weight, diagnoses, allergies, and other patient-specific information. Each field is populated with data that define that attribute (e.g., Smith, female, 64 years, 62 inches, 125 pounds, etc.). To keep track of the drugs that Ms. Smith is taking, we could include fields for all features of the drugs in the patient table. However, this would be inefficient to maintain the data because the recommended dosage, route of administration, etc. are the same for many patients, so the table would have a large amount of redundant data. Instead, a separate drug table with fields for drug name, dose, cost, etc. should be maintained.

For efficiency and compactness, there should be no redundant or duplicate data. The database parlance for this situation is to say that the database is "highly normalized." Additionally, each record must be identified uniquely by one or more attributes because the database must be able to distinguish each patient individually. If some patients had the same last name or the same diagnosis, there might be a chance of confusing the patient records. Every table should have a field that contains unique data for each patient. This unique identifier could be a social security number because no two people should have the same social security number, or in the interest of privacy, each record might include a field for a patient record number. This unique identifier field is called the primary key.

The relationships between tables allow users to extract information from a variety of tables through a query that involves a "join" operation, which allows the user to pull together information from fields in different tables. This is possible because the unique identifier fields in each table (called the primary and foreign keys) are linked. For example, joining patient name and drug name would generate a list of all patients, the drugs they are taking, and their associated dosage information. In performing a query, the user can search all fields of the database or select specific fields. In a large and complex system such as an electronic health record, the relational database consists of many tables, all of which would be linked through primary–foreign key relationships.

4.4.2 Database Searching

The process of specifying what the user is seeking is done through the use of filters, which can take many forms. The filter can be a word or a group of words, a numerical value or range, or a combination of both (e.g., systolic blood pressure over 130). Filters must conform to database controlled vocabularies. By carefully selecting an appropriate filter and specifying which field to search, a user can retrieve the needed information. For example, one could search the diagnosis field for "myocardial infarction." However, someone entering data on a patient might have entered "heart attack" or "coronary artery thrombosis" as alternative terms for the same condition. This then becomes a problem if one queries the database to find out how many patients admitted to the hospital were diagnosed as having a myocardial infarction. The records containing heart attack or coronary artery thrombosis would not be retrieved.

4.4.3 Role of Vocabularies

From the preceding example, the importance of controlled vocabularies should be obvious. How can data be retrieved and analyzed if they are not entered consistently? In this hypothetical case, the controlled vocabulary represents an agreement to use a single term to identify the disease. If, at some later date, it becomes necessary to compare data from both the inpatient and outpatient databases, the task will be much easier if both databases use the same vocabulary. Ideally, only one database would be used for the whole institution and it would use a consistent vocabulary. This situation is rarely the case, and the reconciliation of data across multiple institutional databases has become a major challenge to the future of efficient and cost-effective healthcare provision (see Chapters 16 and 18 for more detail).

To facilitate the entry of the correct terms into the database, data are often entered using a form (or menu on a computer screen) on which, for example, an enumerated list of conditions is provided and the person entering the data picks from the list. Two potential problems exist with this scenario. First, the list of conditions could become very long and the notion of categories and subcategories becomes helpful, as in the discussion of tax-onomies and ontologies later. Second, one of the entries on the enumerated list may not quite fit the clinical situation, so there should be some flexibility in what can be entered. A review of terms that do not fit can lead to further systemization or an extension of the vocabulary. Controlled vocabularies are not only important for patient databases, but also are extremely important to bibliographic databases.

4.5 BIBLIOGRAPHIC DATABASE CONTROLLED VOCABULARIES

4.5.1 Vocabulary Design

Bibliographic databases are tools for finding published materials; for example, library catalogs are bibliographic databases used to find items owned by a library. Each record describes a specific work, such as a book or journal; fields include title, abstract, author, author affiliation, publication date, volume, page numbers, subject headings, etc. Chapter 5 describes a variety of bibliographic databases and how to build effective search strategies for them. This section focuses on one aspect of database searching: the use of controlled vocabularies. It is important to understand how to use these vocabularies because they can make information retrieval more efficient. They are also extremely important in the devel-opment of pharmacy information systems. Although perhaps not consciously aware of it, pharmacists see and use controlled vocabularies in their daily practice whenever they use a bibliographic database or pharmacy information system.

Every field in a bibliographic database is associated with a list (index) of all the data or words in each field. Most electronic databases today have the capability to search all fields for keywords; for example, a search for the term "atorvastatin" in all database fields would retrieve records with that term appearing in any field (e.g., title, abstract, or subject). But many of the documents described in those records would not truly reflect the subject of atorvastatin. For example, an article about hypertension might incidentally mention in its abstract that atorvastatin was prescribed for the patient. A search of all fields for the term atorvastatin would retrieve this record, but the article would not be relevant because it focused on a condition unrelated to the drug of interest.

Users can restrict searches to find their keywords in the index of one particular field (see Chapter 5). It is relatively straightforward to search for information in fields that are well defined, such as author and title; however, fields describing the subjects of articles are more ambiguous because subjects can be described in many ways. As discussed in Section 4.4.2 using the example of myocardial infarction, if a database uses a specific term to describe a subject, then a search of the subject field must use the same term that the database has in the subject field. A search for "atorvastatin" restricted to the subject field would retrieve records that listed atorvastatin only in the subject field and eliminate articles in which it was only briefly mentioned, but was not the subject of the article. This search would yield better results than a search of all fields.

However, the success of the search would depend on what term the database used. If the database used "Lipitor" rather than "atorvastatin," the subject search would be unsuccess-ful. The user would be required to enter synonyms for atorvastatin in the search query in order to locate all records on the subject. When databases employ controlled vocabularies, such issues can be avoided. This section describes how searchers can use controlled vocabularies to focus subject searches and retrieve better results.

4.5.2 Term Mapping

A bibliographic database's controlled vocabulary is a set of standard terms, or descriptors, used to describe subjects. A controlled vocabulary might designate the term "aspirin" as synonymous with the trade name "Ecotrin" and the chemical name "acetylsalicylic acid." Database designers choose preferred terms as descriptors of subjects (e.g., aspirin) and identify synonymous terms that database users might enter in searching for that subject (entry terms). The process by which a database bundles entry terms and points them to preferred terms is called "term mapping." A search of the term "Ecotrin" or "acetylsalicylic acid" in a subject field that used this controlled vocabulary would find records containing "aspirin" in the subject field because the entry terms mapped to aspirin.

Humans use a variety of names, synonyms, abbreviations, acronyms, etc. to describe a single topic. Computers can retrieve records with entry terms in any field, but they require a controlled vocabulary to map the entry terms to the preferred subject terms. For example, as discussed before, a database might use "myocardial infarction" as the preferred term for "heart attack." "Heart attack" is designated as an entry term. A user who did not know the preferred term and limited the search for "heart attack" to the subject field would rely on term mapping to find records with "myocardial infarction" in the subject field. If the entry term were not included in the controlled vocabulary as a synonymous term, it would not map to the preferred subject term and the search would retrieve no results.

Limiting bibliographic database searches to the subject field helps focus searches by excluding records that contain entry terms in a field other than the subject field. In order to limit a search to the subject field of a database effectively, the searcher must use the same terminology as the database in defining the subject; however, it is often difficult to guess the correct database terminology. Some interfaces to bibliographic databases provide messages to users that their entry terms do or do not map to subject headings (e.g., "Did you mean…?").

Some databases include a thesaurus (structured list) of subject headings that users can search to find the preferred terminology and effectively search the subject field using the appropriate vocabulary. Other databases that use a controlled vocabulary do not include a thesaurus. In these cases, the subject search becomes an iterative process in which the user first searches all fields, finds a few relevant results, identifies index terms associated with those records, and launches new searches for those terms in the subject field.

Some databases have no controlled vocabulary, so there is no facilitated subject search capability; the database searches every field for the terms exactly as entered. This is called "free text" searching. Humans understand synonymy (many terms with the same meaning) and ambiguity (many meanings of one term) in language. We intuitively include synonyms in describing a term, and we understand the meaning of terms by their context. A database without a controlled vocabulary cannot manage synonymy or ambiguity; free-text searches should include synonyms, and database search tools should be used to define terms as precisely as possible. Chapter 5 discusses database search tools.

Controlled vocabularies help manage synonomy through term mapping, decreasing the need for users to include all synonyms in subject searches. However, remember that good term mapping is dependent on the database designers' inclusion of all synonyms in the vocabulary. Also, synonomy of terms depends on the context of the database. A bibliographic database might define the generic and trade names of drugs as synonymous for the purpose of defining subjects of articles, but a formulary database of drug products might not because brand-name drugs and generic drugs are different products, possibly with different formulations and prices.

Controlled vocabularies also help manage ambiguity. For example, the term "bridge" could be used to describe a piece of dental work, a structure that helps people cross a river, or a card game. The biomedical literature database, MEDLINE, uses the preferred term, "denture, partial, fixed," to describe the concept of dental bridges, and it includes the term "bridge" as an entry (i.e., synonymous) term. A search of the *subject field* for the term "bridge" would retrieve records containing "denture, partial, fixed" as a subject and exclude the large number of records that contain the word "bridge" in an entirely different context. For example, it is likely that a MEDLINE search of *all fields* for "bridge" would retrieve a large number of records that refer to molecular bridges or records that contain the phrase, "bridge the gap" in the title or abstract. Using the controlled vocabulary helps ensure that the searcher uses the same language the database uses to describe the subject.

4.5.3 Medical Subject Headings

In the biomedical literature database, MEDLINE, the preferred controlled vocabulary terms of the subject field are called medical subject headings (MeSH). MEDLINE includes a thesaurus, the MeSH database, to help users locate preferred MeSH terms for their subjects. MeSH evolved over 150 years.[1] Long before computers and electronic databases, controlled vocabularies were used to search the biomedical literature.

The idea that every article in the medical literature should be tagged with subject headings and indexed under those subjects was conceived in the late nineteenth century by John Shaw Billings, assistant to the U.S. Surgeon General, who collected and stored the medical literature of the time. In 1874, he began indexing the collection by author and subject and in 1880 produced the "Index-Catalogue of the Library of the Surgeon-General." Because of the long delay from publication of an article to the article being included in the index catalogue, Billings started publishing monthly current awareness updates of the index, called *Index Medicus*.

Billings and his successors continued the work of indexing medical literature until 1956, when Congress named this collection the National Library of Medicine (NLM) and made it part of the U.S. Public Health Service (later the National Institutes of Health). In 1960, the NLM produced a list of standardized subject headings, the first edition of MeSH. The only way to search the biomedical literature by subject at that time was by using MeSH. Users looked for entry terms in print thesauruses to find appropriate MeSH terms to search subject indexes.

The first computer-searchable database of medical literature, called the Medical Literature Analysis and Retrieval System (MEDLARS), was produced in 1964. Part of this system was the MeSH database, which was the first version of MeSH to be organized in a hierarchy with "broader than" and "narrower than" relationships. At that time, searches could be submitted to the NLM, where they were formulated and entered into a computer via punched paper cards. Turnaround time for a search request was 4–6 weeks! In 1971, NLM introduced MEDLARS Online (MEDLINE), which could be searched via telecommunications networks. Users still were required to use MeSH terms to search the subject field. MEDLINE evolved with computer technology; it has been offered in a variety of formats such as CD-ROMs and stand-alone database packages.

As the Internet developed, searching became more user friendly, databases became accessible online, and free-text searching became possible. Searchers can now find their keywords in any field of MEDLINE. Free-text searching is very powerful; however, using the MeSH controlled vocabulary helps to focus searches in bibliographic databases and retrieve more relevant results. Chapters 5 and 11 discuss free-text searching in the Web environment. In 1997, the NLM produced PubMed, the first free, online version of MEDLINE. Currently, over 18 million records, each corresponding to a specific article, are contained in PubMed, and MeSH has over 25,000 subject headings. In 2009, approximately 50,000 new research articles were indexed and placed into PubMed each month.

4.5.4 Components and Organization of the MEDLINE Controlled Vocabulary

The main components of the MEDLINE controlled vocabulary are the MeSH terms. NLM designates MeSH terms as preferred terms to describe concepts. The MeSH terms have associated entry terms that can be mapped to MeSH terms. MeSH also includes qualifiers, which are subheadings that can be attached to MeSH terms. These are specific, standardized terms. Approximately 80 subheadings are available and no single MeSH term can be associated with all 80 subheadings. Subheadings are very useful in more precisely defining search terms and thereby focusing a search. For example, a MeSH term describing a disease or condition can be associated with subheadings such as "drug therapy" and a MeSH term describing a drug can be associated with subheadings such as "therapeutic use."

In addition to the two just mentioned, several subheadings are particularly useful to pharmacists. For diseases and conditions, "chemically induced" is a powerful subheading to use when searching for a disease or condition that resulted from the use of a drug. For drugs, the many useful subheadings include "administration and dosage," "adverse effects," "classification," "contraindications," "pharmacokinetics," "pharmacology," and more. Searchers always have the option to include more than one concept in their search query so that a drug name and the term "adverse effects" could be searched together ("ANDed") to retrieve all the records containing both terms. However, using the attached subheading ensures that the term will be associated with the main subject heading and not used in another unrelated context.

A third component of the MEDLINE controlled vocabulary is the substance index. Separate from the MeSH index, it was developed because of the need to index many new drug names. The drug names are "supplementary concepts" analogous to MeSH terms. Older drug names usually appear as MeSH terms because they were indexed before the supplementary concepts were implemented; newer drug names are usually indexed only as

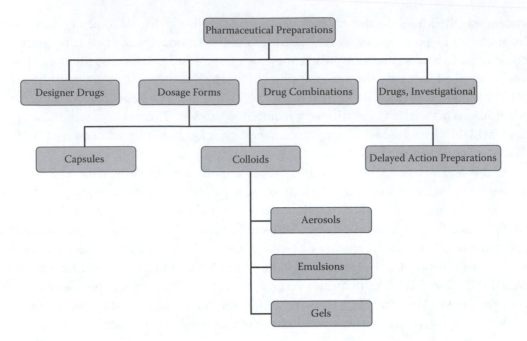

FIGURE 4.1   MeSH hierarchy.

supplementary concepts. Entry terms will map to both the MeSH and substance indexes, but supplementary concepts do not currently have subheadings associated with them. When an individual searches a drug name that is indexed as a supplementary concept, the searcher must enter the subheading as a separate term.

The MeSH controlled vocabulary is a taxonomy (see Section 4.7) in which terms are organized in a hierarchy or "tree" structure. Each term in the taxonomy is related to other terms as either broader or narrower. Users can browse for terms in the MeSH database and see the hierarchical arrangement with broader and narrower terms.

Figure 4.1 shows an example of a tree structure in MeSH. The MeSH term for drugs is "pharmaceutical preparations," and narrower terms fall below this term. The hierarchy conveys meaning and context. For example, gels are a type of colloid, which is a dosage form. MeSH is polyhierarchical, meaning that a subject heading (MeSH term) can appear in more than one category (hierarchy) because it logically falls into more than one category. For example, "colloids" falls under "dosage forms" in the "chemicals and drugs" branch of the tree, as well as under "chemistry, physical" in the physical sciences branch because not all colloids are drugs.

There are other relationships between MeSH terms that are associative. For example, the terms "anticarcinogenic agents" and "antineoplastic agents" are associated, but not exactly synonymous; therefore, each is a unique concept (i.e., MeSH term), rather than one being a preferred term and the other being a synonym. With these associative relationships, users obtain cross-references in the MeSH database. A search for "antineoplastic agents" shows the position of the term in the hierarchy, and it provides the cross-reference "see also anticarcinogenic agents." Some vocabularies have additional types of relationships between concepts so that they form an ontology (see Section 4.8).

4.6 DRUG NOMENCLATURE STANDARDS

4.6.1 Drug Names

Figure 4.2 illustrates some of the many names a drug can have; these nomenclature standards are all examples of controlled vocabularies. Investigational drug codes are often used for drugs early in development, when chemical names are trade secrets. These codes sometimes appear in the literature at early stages. The chemical abstracts registry number (CAS RN) is a unique identifier for chemical compounds. Many bibliographic databases include a CAS RN field that helps specify drug searches. Generic names are assigned later in a drug's development by the United States Adopted Names (USAN) Council.[2] Sometimes international (INN), British (BAN), or Japanese (JAN) adopted names differ from the USAN name, so a comprehensive search would have to include all adopted names.

National drug codes (NDCs) are designations assigned to drug products distributed in the United States.[3] As introduced in Section 4.3, NDC codes have three parts. The first part, the labeler code, is assigned by the Food and Drug Administration. A labeler is any company that manufactures, repackages, or distributes a drug product. The second part, the product code, identifies a specific strength, dosage form, and formulation. The third part, the package code, identifies the package size. Both the product and package codes are assigned by the manufacturer and, unfortunately, are not necessarily consistent or permanent.

Drugs are not usually identified by NDCs in the literature, but NDCs are often used in pharmacy information systems as a standard vocabulary that allows different parts of the system to communicate. The choice of appropriate drug nomenclature depends on where the drug is in its development and what database is being searched[4] (see Figure 5.5 in Chapter 5). For example, the most effective way to search for a drug in a hospital system may be the NDC; some chemical literature databases can be searched very effectively by chemical structure; and the best nomenclature to use in MEDLINE is the generic (USAN) name.

Generic Name:	Ibuprofen
USAN, INN, BAN, JAN:	Ibuprofen
Research Code Designation:	U-18, 573
CAS Registry Number:	15687-27-1
Chemical Names:	Benzeneacetic acid, α-methyl-4-2(methylpropyl) (±)- (±) p-Isobutylhydratropic acid (±)-2-(p-Isobutylphenyl) propionic acid
Proprietary Names:	Advil (Whitehall-Robins) Midol 200 (Sterling Health U.S.A.) Motrin (Pharmacia and Upjohn) Nuprin Caplets and Tablets (Bristol-Meyers Products)
Molecular Formula:	$C_{13}H_{18}O_2$
Structural Formula:	

FIGURE 4.2   Names of ibuprofen.

4.6.2 The Unified Medical Language System

A major barrier to system interoperability is that different information systems use different vocabularies. The unified medical language system (UMLS) is a semantic network of bibliographic database vocabularies; it is a metathesaurus (a thesaurus of thesauruses) including over 60 vocabularies used in the biomedical sciences. The NLM established the UMLS to facilitate interoperability between information systems.[5] One goal of the UMLS project is to unify many disparate vocabularies. All words and phrases that have the same meaning are grouped under a concept name. Each concept name has a unique code. All terms that have the same meaning are associated with one concept and assigned the same code. If a patient record system and a pharmacy ordering system displayed different names for a drug, but the two names were associated with the same concept code, the systems would know that the two designations referred to the same drug

The UMLS includes MeSH, SNOMED, ICD, GO, and many more. It also includes a controlled vocabulary for drug information, RxNorm, which reconciles all the different terms used for drugs and defines synonymy at a more granular level than merely the generic name.[6] Granularity refers to the specificity of the preferred terms for concepts. The names for drugs in RxNorm are clinical drug names, which include ingredients, dosage form, and strength. MEDLINE would include 325 mg tablets and 81 mg tablets of aspirin under one concept of aspirin; however, RxNorm would define these as unique concepts.

RxNorm also includes unique concepts for brand names. It provides this granular representation of clinical drugs in a semantic network with concepts linked by a variety of relationships. Figure 4.3 shows concepts related to the clinical drug "cetirizine 5 mg oral tablet" and the semantic network in which they are related to brand names, ingredients, etc. The goal of RxNorm is to provide the structure needed to support automated clinical processes.

4.7 TAXONOMIES

Although it is already inherent in some of the vocabularies discussed (most notably MeSH), the idea of relationships between terms in a controlled vocabulary is perhaps best illustrated by that of a species taxonomy, a system consisting not only of specific terms, but also of relationships in the form of a lineage representing an evolutionary origin. Thus, *Homo sapiens* is not only a controlled vocabulary term, but it is also part of a lineage (according to the NCBI taxonomy; http://www.ncbi.nlm.nih.gov/Taxonomy/) as follows:

Lineage (full)

cellular organisms; Eukaryota; Fungi/Metazoa group; Metazoa; Eumetazoa; Bilateria; Coelomata; Deuterostomia; Chordata; Craniata; Vertebrata; Gnathostomata; Teleostomi; Euteleostomi; Sarcopterygii; Tetrapoda; Amniota; Mammalia; Theria; Eutheria; Euarchontoglires; Primates; Haplorrhini; Simiiformes; Catarrhini; Hominoidea; Hominidae; Homininae; *Homo*

At each point in the human lineage, a branching leads to the tree of life with humans at the outer leaf. Such a hierarchy is very useful for bringing order out of chaos and takes

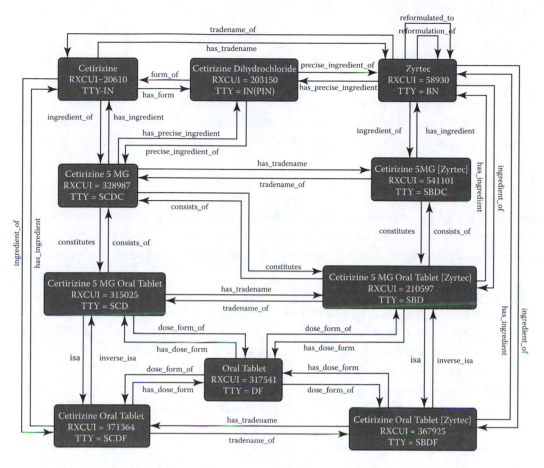

FIGURE 4.3   RxNorm—representation of a medication.

the idea of a vocabulary to the next step of organization, as with the MeSH tree structure. An example of how a taxonomy can be used is illustrated for the RCSB Protein Data Bank (PDB) at www.pdb.org/pdb/browse/browse.do?t=1&useMenu=no. Navigating up and down the taxonomic tree and mousing over each branch indicate how many protein structures have been solved from that species. This is an example of browsing using the hierarchical vocabulary as the tool to locate the level of required detail.

Another similar hierarchical scheme is that developed by the Enzyme Commission for referring to all characterized enzymes (www.chem.qmul.ac.uk/iubmb/enzyme/). This taxonomy is a four-character numeric scheme with each number separated by a period. The following are at the top level in accordance with their most basic enzymatic functions:

1. oxidoreductases;

2. transferases;

3. hydrolases;

4. lyases;

5. isomerases; and

6. ligase.

Again, using the PDB to illustrate this taxonomy, one can navigate a hierarchical tree and review how many structures exist at each level of the enzyme classification. For example, many protein kinases, which are important drug receptors, fall under 2.7.1 (transferases, transferring a phosphorus-containing group, with an alcohol group as the acceptor).

4.8 ONTOLOGIES

Ontologies take the notion of relationships between the terms in a controlled vocabulary even further and have become an essential part of the life sciences over the past few years. There are many definitions of ontology. For our purposes, an ontology describes, in a computer-usable form, the knowledge of a particular domain, including both syntax and semantics.

Perhaps the most well known ontology—one important to the pharmaceutical sciences— is the gene ontology (GO) used to describe gene products. Earlier in the development of molecular biology, as new genes were discovered, there was a tendency to name them after the person who discovered them or to assign a variety of common names. For example, intracellular homology domain of FAS (Apo-1) and the TNF-receptor became known as the death domain. Although experts might be familiar with these various names, someone casually reading in the field would find the task of recognition to be daunting.

Recognition became even more daunting as data on these genes were maintained in a variety of databases, including those on the model organisms studied by many researchers. Organism database examples are *Caenorhabditis elegans* (a nematode worm), *Drosophila melanogaster* (a fruit fly), and *Saccharomyces cerevisiae* (a yeast). Scientists from these databases formed the Gene Ontology Consortium to ensure that consistent names and organization were used across the different databases. Therefore, a researcher could be assured that if he or she referred to a gene in one database, the gene with the same name in another database was undoubtedly related (a homolog). The gene ontology is actually three ontologies that cover cellular location, biological function, and molecular process. Just like in a taxonomy, relationships are specified; however, the gene ontology is more complicated because a given term can derive from multiple parents. For example, a transmembrane receptor tyrosine protein kinase is a transmembrane receptor *and* a protein tyrosine kinase.

In some cases, attempts are made to capture the complete knowledge of a domain within an ontology and to build a database directly from that ontology. A complete discussion is beyond the scope of this introductory textbook, but be aware that ontologies play a vital part in the organization of information that pharmacists and pharmaceutical scientists are likely to encounter. RxNorm is an example of an ontology that may someday be used in hospital and pharmacy information systems because these systems need to have drug vocabularies with relationships between terms that are more complex than a hierarchical, broader/narrower relationship.

4.9 THE SEMANTIC WEB

The World Wide Web is a natural part of our lives, but why does it work so well? At the heart of the Web is the notion of a hyperlink that links two pieces of information, making it simple to navigate and follow a trail of inquiry. This is a powerful tool. To date, many of these links are generated from human knowledge: A human decides to link a term on one Web page to another page. Although this is useful, the human needs to have discovered that remote page and decided that it is worth linking to. The human may not have discovered many more relevant pages.

What if the most relevant pages could be discovered automatically? How could this happen? If terms were used consistently, the most relevant pages might be those that contain the most references to those terms. The terms would need to be semantically consistent to be discovered. That is, not only would the word or phrase need to be found, but also the context in which that word or phrase is used would need to be the same. This process is not trivial; however, automatic discovery based on semantic similarity is one feature of what is called the semantic Web, also referred to as Web 3.0.

The idea of the semantic Web is broader than mapping words and phrases that are contextually the same across multiple Web pages and Web sites. It extends to the common representation of data as well. Consider online life today. It is likely that an individual uses an online calendar tool and a tool to manage photographs, as well as to conduct online banking. Today, different applications handle each task, yet some of the data in each application overlap. I should be able to pull up my calendar and see the photographs I took on a particular day and the bank transactions I did that day; for that matter, I should be able to review e-mail that I flagged important that day, PowerPoint slides I made that day, and so on.

For this scenario to be possible, the common data that tie these applications together must be represented in the same way (i.e., be semantically consistent), and the applications have to work together by using a common application program interface (API) that recognizes these semantics. This day is coming and will be of great importance to pharmacy informatics through integrating many of the online applications discussed in the coming pages. At the heart of this revolution is the use of controlled vocabularies.

4.10 A FINAL THOUGHT

We began this chapter by introducing the *Oxford English Dictionary* as an example of a controlled vocabulary—in this case, for the complete English language. The dictionary has an interesting history (www.oed.com) that is worth considering. Essentially, it was assembled, over a longer time than anyone could have imagined, by a number of volunteers who wrote draft definitions for words in the dictionary. The interesting part is that a major contributor, W. C. Minor, was an inmate of the Broadmoor Asylum for the Criminally Insane.

Today, knowledge is assembled in a not dissimilar way (minus most of the insanity) through Wikipedia in what is now known as a "wisdom-of-crowds" approach. This approach is not subject to controlled vocabularies, but the increased power that could be achieved if it were should not go unnoticed. We believe the day will come when the full power of semantic consistency will be realized. If it were, the reader could now be reading a document and mousing over the words to reveal definitions of words and phrases that were automatically obtained from the most relevant source, thereby enhancing the learning experience.

4.11 CONCLUSION

Although standards and controlled vocabularies might initially seem to be dry and arcane topics, they form the underpinnings of much of pharmacy and medical informatics. This textbook describes a number of systems and processes, such as the electronic health record, pharmacy information systems, clinical decision support systems, electronic prescribing, and bar coding, that often operate independently and use different vocabularies. The hope is that the effective use of standard vocabularies by all enterprises will increase interoperability across systems and ultimately decrease medication errors and improve patient safety and health outcomes.

Pharmacists who understand the concepts of the controlled vocabularies and standards and their implications will be much more sophisticated and successful users of informatics systems. For those interested in a career in pharmacy informatics, intimate knowledge of vocabularies and standards is essential.

REFERENCES

1. Knoben, J. E., Phillips, S. J., and Szczur, M. R. The National Library of Medicine and drug information. Part I: Present resources. *Drug Information Journal* 2004. 38:69–81.
2. United States Adopted Names Council. *USP dictionary of USAN and international drug names.* Rockville, MD: United States Pharmacopeia Convention, 2007.
3. U.S. FDA Center for Drug Evaluation and Research. *The national drug code directory* (available at http://www.fda.gov/cder/ndc/). Silver Spring, MD: U.S. Food and Drug Administration, 2009.
4. Snow, B. *Drug information: A guide to current resources,* 3rd ed. New York: Neal Schuman, 2008.
5. Bodenreider, O. The unified medical language system (UMLS): Integrating biomedical terminology. *Nucleic Acids Research* 2004. 32:267–270.
6. Liu, S., Ma, W., Moore, R., et al. RxNorm: Prescription for electronic drug information exchange. *IT Pro* 2005. 7:17–23.

CHAPTER 5

Literature and the World Wide Web

Susan M. McGuinness

CONTENTS

5.1 INTRODUCTION: WHY STUDY LITERATURE AND WEB SEARCHING?

In today's information explosion, all professionals must possess and continually hone fundamental information skills in order to maintain standards of excellence in their work. Pharmacists, as partners in the healthcare team environment, rely on accurate, timely drug information in every phase of clinical decision making, including choice of drug, appropriate dosage and administration, safety, cost effectiveness, adverse effects, drug interactions, and more. Pharmacists are responsible for effectively communicating this information to other healthcare providers, patients and their families, and the public. Mastering information competencies helps develop critical thinking skills needed throughout pharmacy students' education and careers.

The Accreditation Council for Pharmaceutical Education (ACPE) standards for the doctor of pharmacy (Pharm.D.) curriculum state that graduates must be able to "retrieve, analyze, and interpret the professional, lay, and scientific literature to provide drug information and counseling to patients, their families or care givers, and other involved health care providers [and] demonstrate expertise in informatics" in order to practice pharmacy. The ACPE standards also require that information resources be integrated into teaching programs and that schools "should provide organized programs to teach faculty, preceptors, and students the effective and efficient use of the library and educational resources."[1] The Association of College and Research Libraries (ACRL) adopted the term "information literacy" to define broad information competency standards applicable to all fields of higher education.[2] The ACRL standards include performance indicators for identifying information needs including accessing, evaluating, and using information appropriately.

Pharmacists play an important role in patient care and safety. Recent reviews of the literature show that pharmacist interventions have a positive impact on patient outcomes.[3–6] In order to make evidence-based decisions about patient care, pharmacists must know how to access and evaluate the current medical literature.[7] Database and Internet search skills are essential, yet these skills are sometimes lacking.

A study of pharmacists' computer skills found that database and Internet search skills were needed to practice effectively.[8] In an information needs assessment survey (unpublished), UCSD Pharm.D. students reported that during their fourth-year clinical rotations (advanced pharmacy practice experiences; APPEs), questions directly related to patient care arose more than once a day. Their information needs encompass a wide range of topics. For example, they may need to learn about a disease etiology to prepare for rounds or find the appropriate drug dosage quickly for a particular patient. This requires knowledge of the best sources of background information, numerical data (e.g., dosage limits for patients with compromised renal function), and primary literature about diseases, conditions, and therapies, as well as the skills to use information sources effectively.

Pharmacists who consult with patients about their conditions and medications need to be aware of appropriate sources of patient information and be able to ensure that patients understand the information. (Consumer health literacy and drug information sources for consumers are discussed in Chapter 11.) The ability to identify the best source and locate the best evidence to answer a clinical question is one of the most important skills the pharmacist needs. Some pharmacists specialize in drug information,[9] but the information search, retrieval, evaluation, and management skills described in this chapter are essential competencies for all pharmacists, whatever their chosen specialization.

5.2 TYPES OF INFORMATION SOURCES

The first step in searching is choosing the appropriate source for the topic. Scientific information exists in a variety of sources, which can be divided into three categories: primary, secondary, and tertiary.[10] Primary sources include conference proceedings, patents, dissertations, articles reporting the results of clinical trials, and other publications of original research. It is difficult to keep up with primary sources directly, given the immense volume of new information that is constantly generated.

Tertiary sources include information that has been compiled and repackaged, such as meta-analyses, practice guidelines, review articles, textbooks, encyclopedias, and Web sites. Some databases that provide summary information from reviews of primary literature also fit in the category of tertiary sources. For example, Clinical Pharmacology and Micromedex are drug information databases to which many hospitals and academic institutions subscribe. Pharmacists should be familiar with commercially available tertiary drug information databases and be able to evaluate and select the best database products to support the services they provide. (Chapter 11 describes specific tertiary sources of drug information in detail.)

Secondary sources are indexes that provide information about and access to primary and tertiary sources. The index of a textbook is an example of a secondary source that points the reader to specific information within the book. Online indexes created and maintained by abstracting and indexing services organize information about articles published in newspapers, magazines, scholarly journals, and more. These online indexes are also known as "bibliographic" databases.

For example, the National Library of Medicine (NLM) indexes thousands of biomedical journals and organizes data about the articles in the MEDLINE database. A journal article describing the efficacy of a drug in reducing the incidence of stroke is a primary source that could be found by searching for the term "stroke" in the secondary source, MEDLINE. A library catalog—an index that can be used to find books and other materials owned by the library—is another secondary source. An Internet search engine is also a secondary source because it leads users to both primary and tertiary sources on the Web. Without the searchable secondary source, the primary article would be almost impossible to locate; the searcher would have to browse through thousands of journals in the hope of finding articles about the chosen topic.

The first step to take, when an information need arises, is to decide which type of source would best answer the question. For example, a pharmacy resident who wants background information on the condition of a patient should start with tertiary sources, such as textbooks or Web sites, to find overviews of the subject and would then use a library catalog or Web search engine to locate those tertiary sources. A pharmacist who wants to find the best current evidence on the efficacy of one drug over another could use MEDLINE or another index of medical literature to find primary studies comparing the drugs.

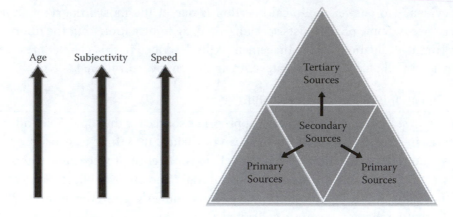

FIGURE 5.1   Types of information sources.

Figure 5.1 depicts a pyramid with primary sources represented at the base to indicate the large volume of these materials; secondary sources, leading users to other sources, are shown at the center, and tertiary sources are at the top of the pyramid. Tertiary sources offer the advantage of providing faster access to information than primary sources do. For example, a pharmacist who needed to find adverse reactions associated with a drug would not begin by searching the primary literature, but would quickly locate the answer in a drug reference book, online drug information database, or Web site. It is important to note, however, that the information found in such tertiary sources is older because of the time lag between the original publication and the development of the repackaged product. It may also be more subjective than information found in primary sources because of potential biases introduced in the review process.

5.3 BIBLIOGRAPHIC DATABASES

A bibliographic database is an electronic index of literature that can include books, journals, magazines, etc. When primary sources are needed, searchers often start with an Internet search engine (e.g., Google or Yahoo!), which scans Web sites and other information on the Web for their search terms. These searches yield a large number of results that must be filtered to select the best results. When the desired sources are journal articles, the Internet search engine can be a useful place to start; for example, Google Scholar indexes the journal articles in PubMed, the NLM's free Web interface to MEDLINE. At the time of this writing, however, bibliographic databases are more powerful and comprehensive search tools for accessing scholarly literature than Internet search engines.

As more journal publishers make their content freely available to the public, this may change; however, because bibliographic databases are specific to subject areas, the selection of the appropriate database provides an initial focus on the subject before the search process even begins. Bibliographic databases also provide better tools for focusing searches to find the most relevant studies while filtering out irrelevant results. For example, in searching for biomedical literature, a PubMed search provides more control over recall and precision than a Google Scholar search.

5.3.1 Bibliographic Database Search Strategies

Thousands of commercially available or free bibliographic databases index popular and scholarly works in a variety of subject areas and physical formats. This chapter focuses on the subset of bibliographic databases that index the biomedical and pharmaceutical literature. Most bibliographic databases used in the sciences are so voluminous that simple keyword searches yield too many results. The challenge of good literature searching is to focus the search precisely on the topic without losing relevant information—that is, to cast a wide enough net to retrieve as many relevant documents as possible, while minimizing the number of irrelevant documents. Information retrieval experts refer to this as a balance between recall and precision, where precision is the ratio of the number of relevant documents to the total number of documents retrieved in a search and recall is the ratio of relevant documents retrieved to the total number of relevant documents that exist.

Because the total number of relevant documents in existence cannot be known, recall can never be measured absolutely, although it can be estimated as it relates to precision. A search that retrieves every document related to the topic would have 100% recall, but would also retrieve a large number of irrelevant documents, or false positives. A search that retrieved no false positives would be 100% precise, but would most likely miss many relevant documents. Recall and precision are inversely proportional and, because relevance is subjective, recall and precision are also subjective. The searcher decides what is relevant.

The searcher must also decide upon the desired precision ratio, which depends on the topic and the intended use of the information. When comprehensive background information is needed, the searcher should aim for high recall and accept lower precision. When learning for the first time about a topic for which there are numerous publications, one can afford to sacrifice total recall for good precision. A good first step in effective literature searching is to consider the possible strategies for searching a topic and to be aware of how those strategies will affect recall and precision.

5.4 GUIDELINES FOR SEARCHING BIBLIOGRAPHIC DATABASES

Each bibliographic database has unique interfaces and search tools with which users must become familiar. This section outlines general guidelines for searching the literature and describes specific strategies for some of the key sources used in the health sciences.

5.4.1 Guideline 1. Identify the Main Concepts

Consider specific aspects of the topic and include terms to help focus the topic. For example, if the topic is diabetes mellitus, consider whether specific age groups, interventions, or other aspects are of interest. (Chapter 11 describes the systematic formulation of clinical questions.)

5.4.2 Guideline 2. Define Search Terms

Think of as many terms as possible that describe the main concepts, including synonyms. This usually entails an iterative process of searching, identifying new search terms, and searching again. If the database offers a controlled vocabulary (see Chapter 4), try to find subject headings or descriptors appropriate to the topic and use those terms to focus the search. Some databases have fields for nomenclatures such as Chemical Abstracts Service (CAS) registry numbers, which can be very useful in searches for a specific chemical substance.

If no controlled vocabulary is available or if the controlled vocabulary is inadequate for the search, it may also be important to include variants of the search terms, such as alternative spellings, plurals, or other forms. Use truncation and "wild cards" to include variants of the search term; for example, a search of the term "formulat*" would retrieve items including the terms "formulate," "formulates," "formulated," "formulation," etc. A search of the term "wom*n" would retrieve items including the terms "woman" and "women." Some databases use the asterisk for both truncation and wild cards; some use different symbols for different functions. The searcher should refer to the help section of each database to determine the correct notation.

5.4.3 Guideline 3. Use Boolean Logic

Boolean logic has many applications; in this context, it is used to combine search terms in a logical query to inform the bibliographic database how to interpret the query and run the search. This helps control recall and precision.

Searching a database for a single term is usually straightforward, but often results in an unmanageable number of hits. Wherever possible, searches should be directed toward specific components of a topic, using a combination of more than one term. For example, searching for "acetaminophen" in MEDLINE will produce thousands of leads. However, searching for "adverse effects of acetaminophen" will limit the results to a more manageable number. When bibliographic databases are searched, the search engine must interpret precisely how the searcher intends to combine the two topics "adverse effects" and "acetaminophen." The way in which two or more topics are combined is expressed with Boolean logic. The three main operators in Boolean logic are "AND," "OR," and "NOT" (designated as "AND NOT" in some databases):

The Boolean "AND" operator will produce results that include all the terms in the search. If we search for "acetaminophen" AND "adverse effects," we would retrieve only the references that discuss both topics (Figure 5.2).

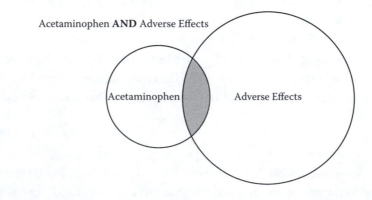

FIGURE 5.2 The Boolean "AND" operator.

Acetaminophen **OR** Adverse Effects

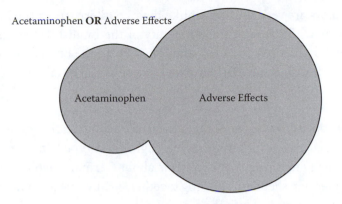

FIGURE 5.3   The Boolean "OR" operator.

The Boolean "OR" operator will retrieve references that contain information about either of the topics alone or in combination with the other. For example, if we search for acetaminophen "OR" adverse effects, we would retrieve all references that address acetaminophen plus all references that address adverse effects of any substance, in addition to adverse effects of acetaminophen (Figure 5.3).

The Boolean "NOT" operator can be quite useful if applied carefully. For example, suppose a search of "acetaminophen" AND "adverse effects" returns a large number of articles reporting adverse effects related to acetaminophen overdose, but the searcher is only interested in adverse effects of standard doses. The Boolean expression, "acetaminophen" AND "adverse effects" NOT "overdose," would retrieve references to articles excluding those that discuss overdose (Figure 5.4). The risk of the NOT operator is that some pertinent articles may be eliminated. For example, articles that discuss adverse effects of acetaminophen with both standard doses and overdose would be eliminated by the NOT operator.

Acetaminophen **AND** Adverse Effects **NOT** Overdose

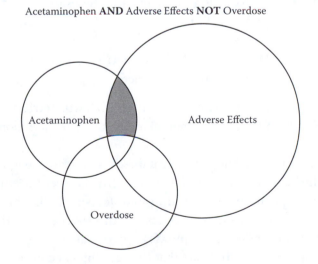

FIGURE 5.4   The Boolean "NOT" operator.

Boolean operators are very specific, unlike their counterparts in the English language: "and," "or," and "not." Because of the ambiguity of the English language, confusing the definitions of the Boolean "AND," "OR," and "NOT" with the English definitions of "and," "or," and "not" can lead to errors. To help avoid this potential pitfall, the following example is offered:

> You plan to adopt a dog and you want to learn more about terriers and German shepherds. If you accessed an Internet search engine and typed "terriers and German shepherds," the search might return items about German shepherd/terrier mixes, terriers and German shepherds playing together, etc. Every hit would have information about both breeds because the search engine interpreted "and" as the Boolean "AND." The "AND" operator will select only those references that discuss both breeds, excluding references with information about only one breed or the other. Some of these results might be relevant to you, if they compare the two breeds, but if you really wanted to read documents about each breed and did not need every document to discuss both breeds, "AND" would produce a set that would be too narrow for your needs. In this case, you should type "terriers OR German shepherds" to get information about either of the breeds.

To summarize the "AND" and "OR" operators, remember that when topics are joined by "AND," the results set will be smaller in number than that referencing both topics individually. The "AND" operator will narrow the target area. The "OR" operator has the opposite effect of expanding the target area and enlarging the number of results.

The ways in which these Boolean operators are invoked differ from one search engine to another. Some systems carry out the operations in left-to-right order and others process the operators in a different sequence; users must consult the help section or advanced search features to learn how the Boolean operators are processed. One can usually override the system's default order of processing by using parentheses to "nest" terms. Operations within parentheses will be performed first:

German shepherd* AND (fox terrier* OR rat terrier*) will retrieve items including German shepherds and either fox or rat terriers.

German shepherd* AND fox terrier* OR rat terrier* will retrieve items about both German shepherds and fox terriers and items about rat terriers alone.

If no operators are employed, then most databases and search engines automatically assume "AND" is implied between multiple terms. This is a very useful default because most searchers of multiple terms do intend to find items including all entered terms. If some combination of AND, OR, and NOT is desired, the searcher must think about intent, entering the operators and nesting the phrases appropriately. Often, Boolean operators must be entered in upper case to indicate that they are intended as operators rather than as English words.

5.4.4 Guideline 4. Use Limits and Qualifiers

Limits and qualifiers (field tags) are used to focus the search. If the set of results is still too large to be useful after a query has been built using synonyms and variants of terms for each concept and the search has been logically ordered using Boolean operators, limits can be applied to the query. Limits are restrictions placed on the query, such as limiting results to a certain date range or document type. Field qualifiers require search terms to be found in a particular field of the database, such as title or author. The general query syntax to use in any bibliographic database is

search term [qualifier] BOOLEAN OPERATOR search term [qualifier]

as in: "White [author] AND Science [journal]."

The field qualifiers in this example are particularly powerful because both "white" and "science" entered as untagged terms would also be found as text words in many documents' title and abstract fields.

Most literature databases include fields for journal title, article title, author, author affiliation, volume, page numbers, subject headings or descriptors, and document type. These fields can be used to focus searches to retrieve only those records with specific data in a given field, such as "review article," "clinical trial," or "patent" in the document-type field. Literature databases usually have more fields and options for limits than Internet search engines; each resource has different searchable fields. Different databases may use different syntax; for example, "white [au]" in one database might be entered as "au = white" in another. Most databases have tools that make it easy to add qualifiers—for example, by selecting from a list—so users usually need not know the syntax.

5.4.5 Guideline 5. Use Multiple Sources

Use multiple databases to broaden the scope of the search. This is especially important when writing literature reviews, which require extensive background information. No one database is comprehensive; it is essential to be aware of the variety in scope and content among sources. Literature on drugs, pharmaceutical sciences, and pharmacy practice may be found in a variety of bibliographic databases. The most appropriate source will depend on the topic, but it is always a good idea to use multiple sources. Figure 5.5 illustrates five bibliographic databases in the context of a drug's life span from discovery to market.

Consider the stage at which the drug was in development for the topic sought. For example, in searching for information about a drug's discovery, the Chemical Abstracts Service (CAS) database may produce better results than the MEDLINE database. There is substantial overlap in the journals indexed by these databases; however, each one has some unique content, is organized differently, and has different search tools. It is therefore important to check multiple sources to capture all the relevant information. Even when two databases that index a similar set of journals are searched, a given search query will often produce different sets of results in each of the databases.

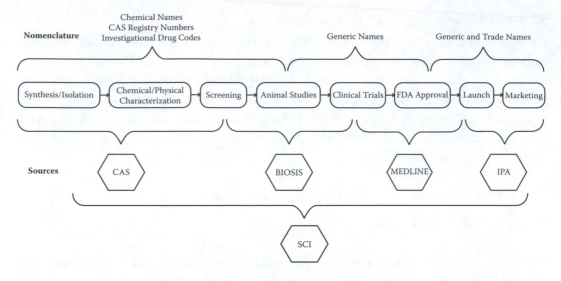

FIGURE 5.5   Drug information resources over the life span of a drug.

5.4.6 Guideline 6. Use Appropriate Drug Nomenclature

Choose appropriate drug nomenclature for the search.[11] As Figure 5.5 illustrates, chemical synthesis is an early stage and marketing is a late stage of drug development. In the early stages, drugs may be designated only by their CAS registry number, chemical name, a manufacturer's investigational drug code, or possibly a generic name. After marketing, a drug will have many more designations, including its *National Drug Codes* and proprietary or brand name. The stage in development informs the decision on vocabulary and helps to select the best database.

The database also influences the decision on appropriate vocabulary. In searching MEDLINE, the medical subject heading (MeSH) term for a drug name (the generic name) is the best choice; in a free-text database, a comprehensive search would include both generic and brand names. Nomenclature can add specificity to a search. For example, the CAS registry number may provide one of the most efficient ways of searching for a drug with multiple names because it uniquely identifies the drug. In this case, it is important to make sure that the database has a designated field for CAS registry numbers. The four databases described in Section 5.5 all have a field for CAS registry number.

5.4.7 Guideline 7. Launch New Searches from Results

Once a list of relevant results has been obtained, look for links to launch new searches from those results and locate additional documents related to the topic. For example, authors listed in references may be hyperlinked, so a click will invoke a new search on that author's name. Look for links to "related articles," which will locate articles with common index terms or common references. Sometimes index terms are listed in database records and can be used to find other records indexed with those same terms.

5.5 BIBLIOGRAPHIC DATABASES FOR PHARMACY AND PHARMACEUTICAL SCIENCES

The databases described next are all searchable by field or across all fields, allow truncation of terms, and use Boolean operators. Some databases use other adjacency operators in addition to the Boolean AND, OR, and NOT. For example, the SAME operator requires all terms to be in the same field. These databases all have the fields one would expect to see in a bibliographic database (e.g., title, journal title, authors, publication type, etc.) and each one has unique fields related to its subject. Users should always check the advanced features and help sections to investigate all options.

5.5.1 The Chemical Abstracts Service

The CAS provides a useful bibliographic database of chemical information, including drug information dating from 1907 to the present. Using the CAS is usually more efficient and effective than searching the biomedical literature when chemical or physical properties of drugs or information on chemical synthesis and product formulations is sought. The CAS database includes several unique fields not typically found in bibliographic databases, such as chemical structures, chemical property data, and manufacturers. The CAS also includes many document types such as patents, dissertations, and technical reports in addition to journal articles. Because of its great breadth, CAS is often used to locate documents not indexed elsewhere. Patents describing the manufacture of a specific drug fall into this category. CAS is available as an online database through several vendors.

SciFinder Scholar (SFS) provides an easily searched interface for CAS and is licensed by many large institutions. SFS provides a wealth of chemical data in addition to citations to published articles and other documents. It has property data, such as pKa and solubility for millions of compounds. A unique and useful feature of SFS is the ability to search by chemical structure through a chemical drawing interface. Users can find chemical reactions by searching for molecules as reactants or products. Although the emphasis of SFS is not primarily on clinical medicine, there is substantial coverage of drugs in research. It is therefore very important to remember this source in order to be comprehensive in searching for drug information outside purely clinical topics.

5.5.2 Biological Abstracts (BIOSIS)

BIOSIS is available as the BIOSIS Previews database through many vendors. It indexes journals, books, book chapters, dissertations, conference proceedings, and patents from 1926 to the present in the major fields of biology, including biochemistry, biotechnology, genetics, nutrition, and many other areas of interest to pharmacists. BIOSIS Previews is useful in finding information about drugs in very early trials in animals. It is also a good source of information from books and book chapters because many bibliographic databases, including MEDLINE, do not include books.

In addition to the fields describing the documents, BIOSIS Previews also includes fields for taxonomic classification, gene names, and sequence data, which provide approaches other than pure subject or keyword searching. BIOSIS Previews uses controlled vocabulary in some of the fields, including one called "concept codes," which are broad subject headings. No thesaurus is available, so users must first run a free-text search and then scan the vocabulary terms in the retrieved records to locate any useful terms to include in a new search (see Chapter 4, Section 4.7.1).

5.5.3 MEDLINE

Created and maintained by the U.S. National Library of Medicine (NLM), MEDLINE is one of the largest literature databases in the health sciences. It indexes close to 5,000 scholarly journals in basic and biomedical sciences, with articles dating from 1950 to the present. MEDLINE does not index conference proceedings or book chapters. The service is available through many vendors who have created various search interfaces. The PubMed interface described in this chapter is freely available on the Web from the NLM and also includes several smaller databases containing some articles that are not part of MEDLINE. The records include fields for the NLM's medical subject headings controlled vocabulary discussed in Chapter 4, as well as chemical names, CAS registry numbers, and a unique identifier number for each record in the database.

When search terms are entered into the main PubMed query box, PubMed searches the fields in a specific order. It first searches the MeSH database, followed by journal titles and then author names. If no match is found, it searches all fields, including article titles and abstracts. PubMed has advanced search features that allow users easily to combine terms with the Boolean AND or OR, as well as to restrict searches to specific fields. There are two ways to search specific fields. The first is to use the "limits" feature and select the desired fields, such as author, journal title, date, or MeSH; the second is to use field qualifiers, which can be entered directly in the query as suffixes to the search terms. For example, if the need was for search terms to be found only in the article title, "your term [ti]" would be entered in the query box.

PubMed also has a very useful field for pharmacologic action MeSH terms. Articles about drugs are assigned MeSH terms for drug names and additional MeSH terms for the drug's pharmacologic action or actions (MeSH[PA]). A search of the MeSH database for a pharmacologic action will yield a MeSH[PA] that links to a list of substances with that pharmacologic action. The PubMed document-type field is especially useful because it indexes over 50 article types to which a search can be limited or that could be excluded from a search (phase I–IV clinical trials, randomized controlled trials, case reports, meta-analyses, practice guidelines, and more). A query could be constructed, for example, to retrieve only randomized, controlled trials or to exclude case reports.

5.5.4 International Pharmaceutical Abstracts

IPA is produced by the American Society of Health-System Pharmacists (ASHP) and covers pharmacy and drug information from 1971 to the present. IPA indexes approximately 800 journals plus conference abstracts from the ASHP, American Pharmacists' Association (APhA), International Pharmaceutical Federation (FIP), and American Association of Colleges of Pharmacy (AACP). Many of the journals indexed in IPA are also indexed in MEDLINE, but a number of journals are unique to IPA, including state and international journals of pharmacy (e.g., *California Journal of Health-System Pharmacy, Irish Pharmacy Journal*) and pharmacy trade publications (e.g., *Drug Topics*), most of which are not indexed in any other database. IPA is a useful supplement to MEDLINE for any search on drug information and is particularly useful for searches involving pharmaceutical formulations, pharmacy practice and business, and other aspects of pharmacy that might not be included in the clinical literature indexed in PubMed.

IPA is available through a number of vendors, and the search methods depend on the vendor. Most interfaces provide an easy way to limit search terms to specific fields and to combine terms with Boolean operators. In addition to the searchable fields typically found in bibliographic databases, IPA includes a field for the therapeutic drug classification of the American Hospital Formulary Service and a descriptor field that uses a controlled vocabulary for subjects.

The four databases described here will very likely remain the core sources for the literature in pharmacy and pharmaceutical sciences, but it is important to keep up to date on the variety of available sources, which will continually change. In summary, be aware of each database's content and know how to use its search tools.

5.6 CITATION SEARCHING

In the process of reviewing scientific literature, it is often desirable to search for papers that discuss or criticize a particular source article of interest in order to gain a better understanding of the original work. Citation searching is the process by which one locates articles that cite a specific source article. Subject indexes such as MEDLINE allow the searcher to find articles that cite the source article, as long as the articles have subject classifications or text in common. But articles associated by subject would not encompass the entire body of documents that refer to the source.

Articles on entirely different subjects might cite the source for a variety of reasons. They might refer to a specific section of the source article (e.g., a laboratory procedure) that does not characterize the broad subject of the work. Citation indexes organize information about journal articles by the articles cited in their reference lists, making citation searching possible. Searchers can easily locate papers that have cited other papers, thus making their literature reviews more comprehensive (Figure 5.6).

The first such index for science, the Science Citation Index (SCI) was conceived in 1955 by Eugene Garfield,[12] who argued that subject searching was only a starting point for a comprehensive review of the literature because subject classifications were not precise enough to capture all of the ideas in a paper. He demonstrated that a high percentage of articles citing a given source were, in fact, associated with a wide variety of subjects. He said that "the scientist is quite often concerned with a particular idea rather than with a complete concept" and described a citation index as an "association-of-ideas" index that would add depth to literature searching. The first SCI, launched in 1965 as print volumes, indexed 600 journals[13]; it now indexes over 3,000.

The ability to examine citation patterns allows researchers to follow the progress of particular fields of study through time. For example, Garfield's citation analysis of the literature in the field of programmed cell death illustrates how new understanding can emerge

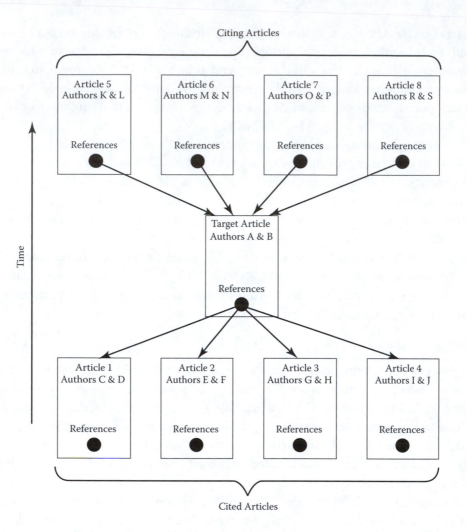

FIGURE 5.6   Organization of articles in a citation index.

from such study.[14] Many tools offering citation searching now exist[15]; two of particular importance to pharmacists are the CAS online database (SFS; see Section 5.5.1) and Google Scholar (see Section 5.7.2). Citation databases enable users to discover subsequent research that follows from older studies, track authors who cite their work, find works that cite the same source articles, identify highly cited authors, and quantitatively measure the impact of a particular article based on the number of citations of that article.

An outgrowth of the SCI is the *Annual SCI Journal Citation Reports* (*JCR*), which ranks journals by citation "impact factor."[13] It is important to be aware of journal impact factors because academic researchers often use them to judge a journal's quality. Impact factors are based on the average number of citations to articles in a given journal. The *JCR* journal impact factor is defined as the average number of citations to articles published in the previous 2 years, divided by the total number of articles published in that journal in that time period. The 2010 issue to the *JCR* will report journal impact factors based on citations in 2010 to articles published in 2008 and 2009.

Journal impact factors have become an established tool for researchers in deciding where to submit manuscripts and for libraries in deciding whether to subscribe to a journal. However, the importance of impact factors in the future is uncertain. The relationship of the impact factor to the actual value of a journal has been a subject of much debate.[16–19] Citation frequency is not the only measure of the value of an article; an article from a clinical practice journal might be heavily used in practice, but unless those practitioners cite the article in a subsequent publication, that use would not be reflected in the impact factor.

Another issue is that journals vary in how they categorize articles; because some article types are not included in the *JCR* calculation of impact factors, it is possible for publishers to inflate the numerator or deflate the denominator by changing how articles are categorized. Also, a journal's accessibility may affect its impact factor as much as or more than its quality. One study of pharmacy-focused journals showed that journals available at no cost to readers tended to have higher impact factors than journals available by subscription only.[20]

Many academic institutions use citation analysis as a measure of the relevance of their research. Using the SCI, one can determine the total number of citations to all the articles by a particular author or group of authors. Another measure that has received attention is Hirsh's "h-index," which quantifies the scientific impact of an individual.[21] An author has h = 10 if he has published 10 articles that have each been cited at least 10 times. As scientific publication evolves in the Web environment, it is easy to predict that research will be disseminated in a variety of new formats and that new measures of relevance, success, or prestige of institutions, authors, and publications will be developed. In the future, journal impact factors will probably carry less weight in evaluating a given publication or author. However, the value of citation searching will likely persist as long as it continues to provide searchers with a unique historical context for scientific research.

5.6.1 The Science Citation Index (SCI)

Before Web-based bibliographic databases emerged, citation searching was painstaking work. To determine the total number of times a paper was cited from the year it was published to the present, the user would consult each annual volume of SCI and add up the number of citing references in each year. Any attempt to repeat the process for the citing articles would quickly become overwhelming. Fortunately, SCI is more easily accessible and citation searching much less arduous today. The online SCI is currently available through several vendors. The Web of Science (WoS) interface to SCI is commonly available at U.S. academic institutions.

Whatever the interface, SCI's records are interlinked through their citations. Searches lead users to references with links to related articles; these articles are related by common citations, rather than by common subjects as in MEDLINE. In addition to keyword searching, SCI offers a citation search function whereby the user can enter an author's name to retrieve a list of that author's publications. Each publication shows the number of citing articles and a link to the full record, which includes lists of cited and citing articles. Users can move backward and forward in time to track the history of an idea from its first communication to the present day. Authors use it to discover how their work is being cited by others. Citations do not imply agreement with the source; the searcher must read the citing article to discover the reason for the citation.

WoS includes the typical searchable fields; its biggest strength is the citation search function. It does not employ a controlled vocabulary, so the results of searches include only records that contain the exact text string that is entered into the search (see Chapter 4). In searching free-text databases like WoS, it is important to remember to use synonyms and variants of search terms to maximize recall, and to include all the concepts related to the topic to maximize precision.

SCI has a broader subject scope than SFS, BIOSIS, MEDLINE, or IPA. It is a good source of literature on computer science, science business, biotechnology, and other subjects beyond biomedical fields. It has no controlled vocabulary, but users can search by keyword in the same way as with other bibliographic databases. The retrieved records include links to cited articles and citing articles (subsequent articles that cite the original article).

5.7 SEARCHING THE WORLD WIDE WEB

The terms "Web" and "Internet" are often used interchangeably, but the Web is actually a subset of the information on the Internet. The Internet consists of infrastructure including transmission devices (routers, switches, computer servers, etc.), interconnecting devices (cables, satellites, etc.), and the data sent through these devices. The Web is the portion of Internet traffic that is intended for display in Web browsers such as Mozilla Firefox or Internet Explorer (words, pictures, movies, sounds, etc.).

Many of the data transmitted over the Internet are not intended for a Web browser; for example, cell phone conversations are part of the Internet, but not part of the Web. Cell phone conversations converted into digital format and transmitted between cell towers are not accessible from Web browsers, though new technologies may be developed to make this possible. Another example of the distinction between the Internet and the Web is seen in the workplace, where data are often stored on a server (i.e., a shared drive) that employees can access from their desktops. This information is part of the Internet but not the Web. The workplace may also have an internal Web site or intranet that is displayed in a Web browser and is part of the Web.

Web information can be subdivided into visible and invisible categories. The visible Web consists of Web pages and data that are accessible through a Web browser. The home page of Google or a video from YouTube are examples of the visible Web. However, some Web information is generated on the fly and intended for use by a single individual—for example, transient Web pages that are dynamically generated from the completion of a form or search.[22] This information is considered part of the invisible Web.

Additionally, limits can be placed on who can and cannot see visible Web pages, thus dividing the visible Web into private and public areas. Journal publishers can make their content available through the visible Web, but much of this content is only available partially or through paid subscription and is thus part of the private Web. Driven by the needs of the public, new technologies to expand the functions of Web browsers will continue to be developed, the Web will continue to grow, and boundaries between Internet and Web, visible and invisible, may begin to disappear.

5.7.1 Web Search Engines

When the Web was first created in 1991, finding information entailed looking for primary and tertiary sources without the aid of a secondary source. One could browse (or "surf") the Web starting with a known Web site (uniform resource locator—URL) and following embedded links to other Web sites. Early Web users quickly accumulated long lists of Web addresses, which led to organizations creating online directories with lists of Web sites organized into broad subject categories. Yahoo! was one of the first such directories.[23] These human-made Web indexes gave users secondary sources to aid them in searching the Web. But even when the Web was still relatively small, the idea of cataloging the entire Web was not only impossible, but also too restrictive for the Web, with its fast-growing and changing content.

Web directories are still excellent starting points for finding information on the Web, but Web search engines now allow users to search the Web for specific keywords of their choice. Automated search engines (e.g., WebCrawler, AltaVista, etc.) began to emerge in 1994 and Yahoo! has since added a search engine to its suite of tools. Search engines use automatic indexing programs, usually called Web crawlers, robots, or spiders, that scan the Web and enter almost every word of text from Web pages and other publicly available documents into an index. They then follow the links within these Web pages and index words from those pages, and so on. Web crawlers also keep track of how words are positioned in the Web documents they scan, so certain words (e.g., words from titles and subtitles) can be given more weight. When using a search engine, the user is not actually searching the entire Web, but rather the search engine's index of the Web.[24]

One reason that different search engines produce different results is that their indexes and their page-ranking methods are different. One study showed that identical queries in MSN Search, Yahoo!, and Google yielded very little overlap when the first page of results was examined.[25] The algorithms that search engines use to locate and rank Web pages are proprietary information unavailable to the public.

5.7.2 Web Search Strategies

Searching the Web for scholarly information is different from searching a bibliographic database; knowing the difference can help the searcher obtain better results. Although it appears to the searcher to be very similar (entering terms into a search form and getting a list of results on that topic), what happens behind the scenes is very different. To understand the difference between searching the Web and searching a bibliographic database, one has to compare the indexes of these sources. As we have discussed, bibliographic databases generate their indexes from defined fields related to the articles (author, title, date, abstract, etc.) rather than the full text of the articles.

Web search engines index almost all words collected from scanning the public Web. This includes the full text of any journal that provides its content to a search engine. Search engines can make agreements with journal publishers to allow their search engine to index the full text of the journal's articles. This does not mean that users can access full text through the search engine, but they can find references to articles that contain search terms anywhere in the article, not only in the searchable fields of a bibliographic database.

Because of their full-text indexing, Web searches generally have a higher recall and lower precision than bibliographic database searches. This can be an advantage in searching for a specific aspect of a topic because relevant search terms might be buried within the text of documents and therefore would not be found in bibliographic database indexes. In such cases, extra effort must be made to increase the precision of the search query to reduce the need to filter through a very large number of results.

Web searches almost always produce more results than bibliographic databases because of the immense volume of information on the Web, including Web sites and other documents not indexed by bibliographic databases. On the other hand, many scholarly documents are not accessible to Web search engines because they reside in private Web space. Search engines do not have agreements with all journal publishers and, at this writing, search engines do not publicize the scope of the journals they do index.

Any discussion of specific Web search strategies will soon be outdated, but some fundamentals may continue into the future. Web searchers are not usually interested in finding every relevant item, and they can usually find what they need on the first page of results because search engines present results in an order of relevance. Searching for the "needle in a haystack" presents a greater challenge. The most important thing to remember in Web searching is that the searcher should imagine the ideal result and be specific in selecting terms that are highly likely to appear in the main section of that ideal result. Try using the Boolean NOT if terms that would be likely to accompany the selected terms but would not constitute an ideal result can be determined.

Search engines will search for exact phrases when quotation marks are used. They ignore certain words, such as "to," "the," "and," "or," etc., and will automatically translate queries of more than one term with the AND operator. Google and Yahoo! require "OR" to be capitalized if it is intended as an operator, and they interpret a minus sign immediately before a term as the Boolean NOT.[26,27] Search engines' Web sites include many strategies about how to get the best results for a searcher's needs. For example, asterisks can be used in a variety of ways as "wild cards"; searchers can even enter a sentence with one or more asterisks placed where they want the search engine to find documents that fill in the missing words. There are numerous additional tips that can be very useful (e.g., placing a tilde [~] before a term in Google requires the search engine to find pages with synonyms to that term). Each search engine offers a wide variety of different tools; although it is not feasible to list them all, it is possible to keep up to date by checking the help pages of the search engine.

Searching the Web for pharmacy information has advantages and disadvantages. As mentioned previously, full-text searching enables one to find the finer details within documents, but it also increases the number of irrelevant results. Because the Web includes documents and sites from unreviewed sources, searchers should carefully evaluate pharmacy and health information found on the Web. (Chapter 11 discusses Web site evaluation criteria for pharmacists and healthcare consumers.) Google Scholar provides a search engine for "scholarly" publications, which include peer-reviewed journals and other works (books, theses, etc.) produced by universities, academic publishers, and professional organizations. It searches the full text of these works, but usually only retrieves a short section of the document highlighting the search term. Members of academic institutions can often access the full text of any work their institution owns by setting preferences in Google Scholar to link to library holdings.

Google Scholar has citation linking similar to SCI, in which documents are linked to citing articles. Documents are also linked to related articles, as in PubMed, where the articles are related by similarity.[28] The disadvantage of Google Scholar is that its users cannot know the scope of scholarly work to which the search engine has access. The scope depends on contractual agreements between Google and a variety of journal and book publishers; Google does not disclose specific information. PubMed and SCI probably index more journals than Google Scholar at this time, but the latter includes additional types of documents, such as books. Therefore, Google will produce a different list of related articles and citing articles than the related articles in PubMed and the citing articles in the SCI.

5.8 INFORMATION MANAGEMENT

Once efficient and effective strategies for searching bibliographic databases and the Web have been determined, it is helpful also to know how to use tools that help organize the information found. Two such tools are alerting services and bibliographic management software. Alerting services, available from most bibliographic databases and journals, provide a means of keeping current with a topic or journal of interest. Bibliographic management software provides users with their individualized bibliographic databases, tools for retrieving information from outside sources, and tools for generating bibliographies.

Most bibliographic databases allow users to save search results in a variety of ways, including printing, e-mailing, and saving text files. They also allow users to save a search strategy, set the database to rerun the search at specific intervals, and send e-mail notifications of new results retrieved since the last search. Search alerts help users maintain current awareness of important topics and save time in the information gathering process.

PubMed provides an excellent alerting service, currently called "My NCBI." Users can run a search in PubMed, save the strategy, and log into their account at any time to run the search again without having to rebuild their query; they can also set up automatic e-mail alerts. The PubMed tool also provides server space where users can save lists of references that they can access from anywhere and can share with others. Most journals provide e-mail alert services that allow users to receive tables of contents to current issues making it easier to keep up with specific journals that are important in everyday practice.

Bibliographic management programs (e.g., EndNote, RefWorks, ProCite, etc.) are valuable tools in organizing the literature read and used. These packages provide a variety of ways to organize this collection of literature, including sorting functions and customizable folders that allow references to be categorized according to topic, project, or any useful grouping. They have sections where users can add notes, links to images, links to the full text document associated with that reference, Web links, etc.

They also have tools for users to search bibliographic databases, such as library catalogs and PubMed, and to import search results directly from those sources. Bibliographic management programs have the capability to display references in a wide variety of styles that conform to the requirements of specific journals and writing style guides (e.g., *Chicago Manual of Style* and the "Uniform Requirements" for manuscripts submitted to biomedical journals). These tools also add functionality to word processing programs whereby users can insert in-text citations and generate reference lists.

Alerting services and bibliographic management tools can help streamline the processes of information gathering, maintaining current awareness, organizing your documents, and appropriately citing the literature in your work.

5.9 CONCLUSION

Current sources of pharmacy and drug information and the search skills to use them effectively are essential in today's environment. Many of the specifics will change over time. Bibliographic databases may merge or change the scope of content they cover, and traditional journals may be replaced by emerging methods of scientific communication. The only truly predictable aspect of the Web is that it will continue to grow exponentially. Google, Yahoo!, and other Web sites offer numerous tools and services in addition to the basic search engines described in this chapter (real-time maps; tools for building Web sites; productivity tools for documents, spreadsheets, and presentations; sharing capabilities; e-mail; social networking functions, etc.).

Imagine the possibilities in the pharmacy practice setting! Already, many pharmacists have the capability to access their pharmacy information systems from mobile devices. The tools will change, but the fundamentals of effective information searching and management will remain constant. Know how to define the topic according to the need, select the appropriate sources, and invest the time to learn how to use the tools those sources offer to retrieve the best information efficiently and effectively.

REFERENCES

1. Accreditation Council for Pharmacy Education. Accreditation standards and guidelines for the professional program leading to the doctor of pharmacy degree. 2007 (available at http://www.acpe-accredit.org/pdf/ACPE_Revised_PharmD_Standards_Adopted_Jan152006.pdf). Chicago, IL: Accreditation Council for Pharmacy Education (accessed January 24, 2009).
2. Association of College and Research Libraries. Information literacy competency standards for higher education. 2008 (available at http://www.ala.org/ala/mgrps/divs/acrl/standards/informationliteracycompetency.cfm). Chicago, IL: American Library Association (accessed January 24, 2009).
3. Machado, M., Bajcar, J., Guzzo, G. C., et al. Sensitivity of patient outcomes to pharmacist interventions. Part II: Systematic review and meta-analysis in hypertension management. *Annals of Pharmacotherapy* 2007. 41:1770–1781.
4. Machado, M., Bajcar, J., Guzzo, G. C., et al. Sensitivity of patient outcomes to pharmacist interventions. Part I: Systematic review and meta-analysis in diabetes management. *Annals of Pharmacotherapy* 2007. 41:1569–1582.
5. Machado, M., Nassor, N., Bajcar, J. M., et al. Sensitivity of patient outcomes to pharmacist interventions. Part III: Systematic review and meta-analysis in hyperlipidemia management. *Annals of Pharmacotherapy* 2008. 42:1195–1207.

6. Wubben, D. P., and Vivian, E. M. Effects of pharmacist outpatient interventions on adults with diabetes mellitus: A systematic review. *Pharmacotherapy* 2008. 28:421–436.

7. Burke, J. M., Miller, W. A., Spencer, A. P., et al. Clinical pharmacist competencies. *Pharmacotherapy* 2008. 28:806–815.

8. Balen, R. M., and Jewesson, P. J. Pharmacist computer skills and needs assessment survey. *Journal of Medical Internet Research* 2004. 6:10.2196/jmir.6.1.e11.

9. Rosenberg, J. M., Koumis, T., Nathan, J. P., et al. Current status of pharmacist-operated drug information centers in the United States. *American Journal of Health-System Pharmacy* 2004. 61:2023–2032.

10. Slaughter, R. L., and Edwards, D. J. *Evaluating drug literature: A statistical approach.* New York: McGraw-Hill, 2001.

11. Snow, B. *Drug information: A guide to current resources,* 3rd ed. New York: Neal Schuman, 2008.

12. Garfield, E. Citation indexes for science; a new dimension in documentation through association of ideas. *Science* 1955. 122:108–111.

13. Garfield, E. Citation Analysis as a tool in journal evaluation—Journals can be ranked by frequency and impact of citations for science policy studies. *Science* 1972. 178:471–479.

14. Garfield, E., and Melino, G. The growth of the cell death field: An analysis from the ISI Science citation index. *Cell Death and Differentiation* 1997. 4:352–361.

15. Roth, D. L. The emergence of competitors to the Science Citation Index and the Web of Science. *Current Science* 2005. 89:1531–1536.

16. Bar-Ilan, J. Informetrics at the beginning of the 21st century—A review. *Journal of Informetrics* 2008. 2:1–52.

17. Cherubini, P. Impact factor fever. *Science* 2008. 322:191.

18. Notkins, A. L. Neutralizing the impact factor culture. *Science* 2008. 322:191.

19. Simons, K. The misused impact factor. *Science* 2008. 322:165.

20. Clauson, K. A., Veronin, M. A., Khanfar, N. M., et al. Open-access publishing for pharmacy-focused journals. *American Journal of Health-System Pharmacy* 2008. 65:1539–1544.

21. Hirsch, J. E. An index to quantify an individual's scientific research output. *Proceedings of the National Academy of Sciences USA* 2005. 102:16569–16572.

22. Devine, L., and Egger-Sider, F. Beyond google: The invisible Web in the academic library. *Journal of Academic Librarianship* 2004. 30:265–269.

23. Northedge, R. Google and beyond: Information retrieval on the World Wide Web. *Indexer* 2007. 25:192–195.

24. Brin, S., and Page, L. The anatomy of a large-scale hypertextual Web search engine. *Computer Networks and ISDN Systems* 1998. 30:107–117.

25. Spink, A., Jansen, B. J., Kathuria, V., et al. Overlap among major Web search engines. *Internet Research* 2006. 16:419–426.

26. Google. Google Web search help. 2009 (available at http://www.google.com/support/web-search). Mountain View, CA: Google, Inc. (accessed April 29, 2009).

27. Yahoo!. Yahoo! search help topics. 2009 (available at http://help.yahoo.com/l/us/yahoo/search/basics/). Sunnyvale, CA: Yahoo!, Inc. (accessed April 29, 2009).

28. Google. Google Scholar help. 2009 (available at http://scholar.google.com/intl/en/scholar/help.html). Mountain View, CA: Google, Inc. (accessed April 29, 2009).

III

Information Systems

CHAPTER 6

Hospital Information Systems

Joshua Lee

CONTENTS

6.1 INTRODUCTION

As we move into the twenty-first century, the meaning of the patient medical record is in flux. What began a century ago as a paper record intended to contain the universe of care provided to a patient in a given setting has transformed into a fluid electronic portal that describes not only local care but also the interactions between health providers and their patients. The record now encompasses the broadest possible definition of data aggregation centered on the patient. This chapter aims to define the evolution of this record from its paper origins to its current state and explain the transition from traditional models of systems generated by healthcare entities to more commercially available products that define the "state-of-the-art" functionality.

To some, the computer has become the equivalent of the stethoscope in healthcare: the ubiquitous, necessary tool that enables providers to embark on their diagnostic and therapeutic journeys. Although it took a while for the stethoscope to reach a peak of adoption and universality, so too will there be a period of adjustment as the electronic health record (EHR) reaches maturity in the marketplace, in the quality of the product, and in the science of correct implementation. As of the writing of this chapter, no consensus exists about the appropriate "lexicon" of functionality of the EHR or about the ideal manner of its adoption.[1,2] EHR implementation is a struggle facing all who specialize in healthcare informatics; it also should spur additional inquiry in this field from all healthcare practitioners invested in improving the current state of medicine in the United States and elsewhere.

One of the key transformations in the EHR is that, in place of static presentation of data for review by healthcare practitioners or even a venue in which to order pharmacologic or diagnostic interventions, the record has also become a place of discourse between healthcare providers. Whether it is messages from a phone call to a physician's office about an appointment or a dialogue between nurse and pharmacist about an optimal drug dosage regimen, the EHR now provides a place of record for these communications and also places them in the correct context of the record.

An even more exciting frontier is the area of patient–provider communications. Many homegrown and vendor-provided EHR products now provide secure Web-based portals for use by patients. These portals allow for secure messaging between patients and clinicians in issues of symptom evaluation, medication queries, and review of results. They also allow patients to keep track of individualized immunization schedules as well as their age-appropriate screening interventions.

6.2 HISTORICAL DEVELOPMENT OF THE ELECTRONIC HEALTH RECORD

The first electronic records were aggregations of observations laid down on paper, either through full narrative entries or scanned copies of notes. The initial term for these records was "electronic medical records" (EMRs) in that they dealt with the record of disease and interventions to cure them. However, as the repository came to include elements of healthcare maintenance and preventative care items, it became a full "health record" spanning the continuum of care, hence becoming an "electronic health record," or EHR. Elements in the following key areas made online records more advantageous[3]:

- *Accessibility.* By making records available through applications distributed on desktops across an enterprise linked to a central server or by using an Internet-based application, patient data became available at any time and nearly any location, without the need for medical records file rooms and file clerks, thereby diminishing the specter of a lost record.

- *Legibility.* The high degree of variability in provider handwriting has often been identified as among the root causes of medication errors.[4,5] The EHR circumvents this by using a typewritten interface and by limiting the kinds of data that are allowable (see Chapter 4). These goals are accomplished by restricting the range of a numeric value to prevent "keying" errors and by making providers choose from a list of allowable values rather than entering variable text.

- *Use of discrete data.* By forcing the use of discrete data elements in representation of the elements of history, medical problems, medications, and even social history documentation, the record allows for aggregation of data across populations and the creation of association with other findings such as laboratory values or radiological findings. A good example of this is the correlation of certain disease states with the appropriate prescribing of classes of medications. Patients with left ventricular heart failure should be prescribed an ACE-inhibitor or angiotensin receptor blocker. If the presence of a left ventricular heart failure is detected but one of these medications is not on the active medication list, the patient's physician can be prompted to prescribe one by sending an electronic alert to him or her.

The first use of EMRs occurred in single institutions that sought to develop systems to support the "business enterprises" of hospitals—namely, the capture of physician orders and their appropriate routing to departments such as laboratory, pharmacy, and radiology and then to processing the associated fees that should be charged for those services. The side benefit of these systems is that they often provided clinicians with access to results (e.g., laboratory and radiology reports) electronically, allowing them to move beyond paper printouts. The manner by which these results were aggregated formed the basis of the first tenet of comprehensive electronic records: an integrated view of patient data across time and specialty.

Individual systems often used proprietary formats to display laboratory and radiology values, which were very useful and tailored for their own systems but were not understandable to a larger interface. Drawing on the experience of early programmers who leveraged the application programming interface (API) and standards initially created by the American Society for Testing and Materials (ASTM), developers learned to create universal messaging protocols to allow for interchange of these data among different computer systems.

Thus, individual hospitals would not be required to do interface programming themselves when they sought to integrate individual departmental systems to their core EMR. This standard, now called health level 7 (HL-7), references the highest level of integration of information and is the industry leader. It is in use at over 1,500 healthcare institutions in the United States.[6] Initially, this allowed providers to see relevant clinical information for a particular inpatient stay. The standard has emerged to allow review of clinical information across time and across modality (see Figure 6.1).

To represent data elements, it also became important to standardize not only the ways in which data were transmitted, but also the way in which each individual data item was rendered to achieve the vision of "discrete" data as identified before. Terminologies were created to control the representation of data, each for its own individual area: diagnoses,

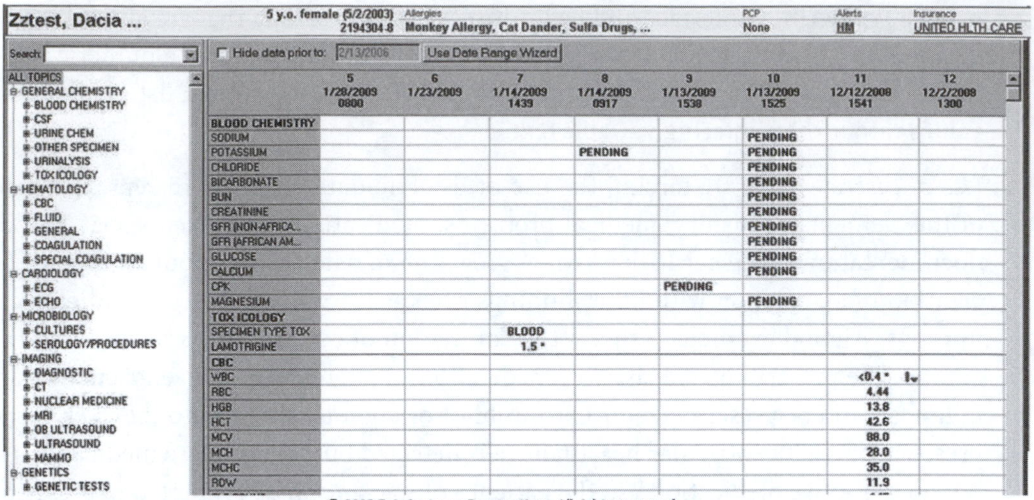

Zztest, Dacia ...	5 y.o. female (5/2/2003) 2194304-8	Allergies Monkey Allergy, Cat Dander, Sulfa Drugs, ...					PCP None	Alerts HM	Insurance UNITED HLTH CARE

Search: ☐ Hide data prior to: 2/13/2006 Use Date Range Wizard

	5 1/28/2009 0800	6 1/23/2009	7 1/14/2009 1439	8 1/14/2009 0917	9 1/13/2009 1538	10 1/13/2009 1525	11 12/12/2008 1541	12 12/2/2008 1300
BLOOD CHEMISTRY								
SODIUM						PENDING		
POTASSIUM				PENDING		PENDING		
CHLORIDE						PENDING		
BICARBONATE						PENDING		
BUN						PENDING		
CREATININE						PENDING		
GFR (NON-AFRICA..						PENDING		
GFR (AFRICAN AM..						PENDING		
GLUCOSE						PENDING		
CALCIUM						PENDING		
CPK					PENDING			
MAGNESIUM						PENDING		
TOXICOLOGY								
SPECIMEN TYPE TOX			BLOOD					
LAMOTRIGINE			1.5 *					
CBC								
WBC							<0.4 *	
RBC							4.44	
HGB							13.8	
HCT							42.6	
MCV							88.0	
MCH							28.0	
MCHC							35.0	
RDW							11.9	

Topics tree (left panel): ALL TOPICS; GENERAL CHEMISTRY — BLOOD CHEMISTRY, CSF, URINE CHEM, OTHER SPECIMEN, URINALYSIS, TOXICOLOGY; HEMATOLOGY — CBC, FLUID, GENERAL, COAGULATION, SPECIAL COAGULATION; CARDIOLOGY — ECG, ECHO; MICROBIOLOGY — CULTURES, SEROLOGY/PROCEDURES; IMAGING — DIAGNOSTIC, CT, NUCLEAR MEDICINE, MRI, OB ULTRASOUND, ULTRASOUND, MAMMO; GENETICS — GENETIC TESTS

FIGURE 6.1   Review of results from many modalities, labs, and radiology across time. (Courtesy of Epic Systems, Madison, Wisconsin.)

procedures, psychiatric diagnoses, clinical observations (e.g., laboratory values, vital signs), and medications. The review of all of these is beyond the scope of this chapter (see Chapter 4 for details), but it is helpful to outline the most common standards used to represent these elements in electronic health records.

6.2.1 Diagnoses

Diagnoses are usually managed by the International Classification of Diseases and its Clinical Modifications (ICD-9-CM), which serves as the "lingua franca" of diagnostic terms in U.S. hospitals. It is used for clinical decision support and for billing support purposes. It is now out of sync with the rest of the world, which has moved to ICD version 10, which is slated to be implemented in the United States by 2013.

To aggregate a group of related diagnoses, the concept of diagnosis-related groups (DRGs) was created; this allows for a smaller number of diagnostic groups to define a given hospitalization, facilitating reimbursement for similar care across hospitals. For example, there are many ICD-9 terms for bacterial pneumonia (e.g., 482.83: pneumonia secondary to Gram-negative organisms, and 482.31: pneumonia secondary to streptococcus). However, many of them are rolled up into larger groups to create a more rational basis for compensating hospitals (e.g., DRG 89: pneumonia with complications, and DRG 90: pneumonia without complications).

6.2.2 Utilization and Procedures

The American Medical Association keeps a master dictionary of procedures (*Current Procedural Terminology*) that encompasses the universe of diagnostic and therapeutic procedures done by providers to patients.[7] Although its use is almost exclusively in the reimbursement realm, it has uses among health services researchers to understand patterns of care.

Furthermore, it is often the way in which requests for procedures (i.e., a laboratory or radiology test) are "ordered" by the core electronic medical record to the recipient ancillary system.

6.2.3 Laboratory Findings and Observations

Although it is not broadly deployed, researchers at the Regenstrief Institute in Indianapolis, Indiana, developed a system of structured data for laboratory findings and later for other observations (e.g., vital signs, electrocardiographic findings). This came to be known as the Logical Observation Identifier Names and Codes (LOINC) terminology.[8]

6.2.4 Nursing Terminologies

To provide structure to nursing documentation at the bedside, a schema of problem (unique to direct bedside care) catalogs of the expected outcomes and the interventions to achieve those outcomes were developed. The most broadly deployed are the North American Nursing Diagnosis Association (NANDA), Nursing Outcomes Classification (NOC), and Nursing Interventions Classification (NIC).[9] However, it is key to note that there is no correlation between the more classical "medical" diagnoses utilized in ICD-9-CM and these nursing diagnoses. Consequently, most inpatient records have two problem lists at any time: those identified by the physician providers and those laid out by the nursing professionals. As more integrated EHRs are implemented, these disparate problem lists will likely need to be harmonized for the sake of interdisciplinary care.

6.2.5 Drug Codes

The challenge in creating structured terminology coding for medications is that the current standard from the U.S. Food and Drug Administration (FDA) is a dictionary entitled the *National Drug Codes* (NDCs) that is driven by the manufacturer and not unique to a specific drug, dose, and route. Rather, it is very much influenced by manufacturer and packaging. Although the National Library of Medicine (NLM) has sought to create a universal standard for transmitting medication information (RxNorm), to date it has not been widely adopted commercially. Most hospital systems rely on commercially prepared, proprietary drug databases with attached clinical decision support information (e.g., checking drug–allergy interaction and drug–drug interaction). The two most common providers in the United States are First Data Bank (San Bruno, California) and Medi-Span (Indianapolis, Indiana).

6.2.6 Implementation of EHRs

Structured terminologies and the manner to link these elements between ancillary systems and the core EMRs could enable institutions and, eventually, vendors to tackle the issue of using EHRs not only to display data, but also to capture observations, notes, and charges and to provide real-time decision support to clinicians. Initial records that emerged from early hospital information systems were pioneered at several academic medical centers in the 1980s and 1990s. Among the most notable were the HELP system developed at the LDS Hospital (Salt Lake City, Utah) and the Regenstrief System developed at Wishard Memorial Hospital (Indianapolis, Indiana). These systems were the first to use the rendering of clinical data, integrated from several systems, to provide physicians with clinical "guidance" in their orders.[10]

The development of these systems was assisted by being completely under the control of their local developers and being implemented in one clinical setting. This allowed for careful tailoring of the work flow to the local customs and resulted in high user adoption. Furthermore, use of the electronic medical record in an academic medical center with a relatively homogenous physician population—all affiliated with that center—also contributed to improved adoption of the record. As recently as 2009, EHR availability and adoption were found to be positively correlated with larger practices.[11]

The key driver behind adoption and the move to the use of vendor-created EHRs has been increased recognition that the healthcare environment is fraught with potential errors and that a systematic approach to care, rather than the individual choice of a given provider, is more likely to result in a beneficial outcome to the patient.[12] This realization led many hospitals and large physician practice groups to consider adoption of electronic records; the composition of most records as well as their track record in improving care will be the next focus of this chapter. (Further discussion of EHR implementation issues can be found in Chapter 18.)

6.3 FUNCTIONAL COMPONENTS OF THE EHR AND THEIR RELATIONSHIP TO OTHER HEALTHCARE SYSTEMS

6.3.1 Integration and Result Review

One key feature of the EHR that distinguishes it from a simple electronic version of a paper record is that it is tightly integrated with other practice management tools that allow for patient identification, patient tracking, and patient financial management. The core underpinning of any record is the master patient index (MPI), which identifies all patients cared for in a given institution and their key demographic characteristics. The MPI is often housed in a registration system that not only informs the core EHR of new patients or changes to the demographics of patients, but also informs all other ancillary systems. When these demographic changes or additions are propagated across hospital or medical group enterprise systems, they are called admission, discharge, and tracking (ADT) messages. The name is a holdover from days when hospital information systems tracked only the locations and characteristics of patients.

These messages have grown more sophisticated over time and now include key financial information relative to that encounter, such as the patient's insurer, prior authorizations in place to cover the care that is to be rendered, and the duration of that authorization. This kind of encounter-based financial information can also be embedded in scheduling systems that are now often linked to EHR systems. These expanded data allow providers to view patient information for individual patients over time and to review findings across an entire clinic schedule day (see Figure 6.2). Additionally, financial information that may inform certain key treatment decisions (e.g., the need to obtain authorization for a given procedure) is facilitated by the transmission of financial and clinical information in the schedule.

Time	Patient	Visit Type	Notes	Status	Visit Type		Signed Tx Pln	Signed Supp Pln	Signed Supp Pln 2
1:00 PM	Assaggio, Training	ONC	Infusion	Scheduled	ONC		n/a	n/a	n/a
1:00 PM	Brigantine, Training	ONC	Infusion	Scheduled	ONC		No	Yes	n/a
1:00 PM	Croce, Training	ONC	Infusion	Scheduled	ONC		n/a	Yes	No
1:00 PM	Donovans, Training	ONC	Infusion	Scheduled	ONC		Yes	Yes	n/a
1:00 PM	Flemings, Training	ONC	Infusion	Scheduled	ONC	ONC	No	Yes	n/a
1:00 PM	Greystone, Training	ONC	Infusion	Scheduled	ONC		n/a	No	Yes
1:00 PM	Humphreys, Training	ONC	Infusion	Scheduled	ONC		No	No	n/a
1:00 PM	Ilfornaio, Training	ONC	Infusion	Scheduled	ONC		n/a	n/a	n/a
1:00 PM	Kaiserhof, Training	ONC	Infusion	Scheduled	ONC		No	n/a	n/a
1:00 PM	Lavache, Training	ONC	Infusion	Scheduled	ONC		Yes	n/a	n/a
1:00 PM	Masala, Training	ONC	Infusion	Scheduled	ONC		Yes	Yes	No
1:00 PM	Nobu, Training	ONC	Infusion	Scheduled	ONC		n/a	n/a	n/a
1:00 PM	Oceanaire, Training	ONC	Infusion	Scheduled	ONC		n/a	n/a	n/a
1:00 PM	Prado, Training	ONC	Infusion	Scheduled	ONC		n/a	n/a	n/a
1:00 PM	Roppongi, Training	ONC	Infusion	Scheduled	ONC		n/a	n/a	n/a
1:00 PM	Sbicca, Training	ONC	Infusion	Scheduled	ONC		n/a	n/a	n/a

MUC ONC CHEMO Department (All Providers)   as of 8.57 PM

FIGURE 6.2   Review of the schedule gives the provider additional information along with the clinical data. (Courtesy of Epic Systems, Madison, Wisconsin.)

The newer generation of EHR products reaches beyond mere textual or numeric data to provide access to images used in the course of care. These radiologic (and other specialty) images are usually maintained in a separate database known as the Picture Archiving and Communication System (PACS), which has its own set of communication and terminology standards, much as do the ones described previously for the EHR.[13] Increasingly, links to such images are embedded within the core EHR so that clinicians, in addition to viewing flow sheets of laboratory data and medication lists, can bring up the image of a recent chest x-ray or computerized tomographic scan in one click. In a manner similar to integration of radiology images, EHRs can also create links to third-party document management systems so that scanned images of documents from outside the institution can be viewed in the patient's record.

However, the management of scanned documents requires very careful indexing and category management so that appropriate documents can be found easily by the casual user. Scanning is an important concern and the work flow of scanning needs to be well established before implementation. Eliminating the paper medical archive by bulk document scanning of historic folders and implementing revised work flows for scanning new documents and the implications thereof have been reviewed.[14]

Newer electronic record systems seek to provide information in ways that support common clinician work flows. For example, in the earlier description of integration with ambulatory or outpatient clinic schedules, the provider needs to see the relevant results from prior *outpatient* encounters or references to the specialty in which the patient is being evaluated that day. On the other hand, when physician providers are conducting rounds in an *inpatient* setting, it is more important to present information relevant to the inpatient

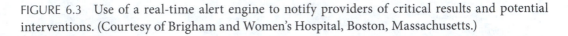

```
   Time: 10/19/94 05:39 AM    Alert#38269              12B phone: x7865
  Alert: DANGEROUSLY LOW SERUM POTASSIUM
 Reason: <BLOOD> K = 3.2 at 04:22 AM, 10/19/94
         Patient is currently on DIGOXIN.

 Relevant medications:                               Alert Details
         LASIX 20 MG IV  BID  Starting on 10/18 <10/17>
         DIGOXIN EVEN days:.125; ODD days:.25 PO          <09/28>

 Actions:
     [ ]A    D/C or EDIT relevant medications
     [ ]B    Order POTASSIUM CHLORIDE IV
     [ ]C    Order POTASSIUM CHLORIDE PO
     [ ]D    Order set: STAT EKG
     [ ]E    Order set: STAT K


                        Bp#2710 was paged on 05:40 AM  Oct 19, 1994
 Covering M.D.:                    Bp#5336     pAge M.D.
     dOne       <done, Go to OE>    coMments      Logic
 Press ALT-O or ALT-G to exit and acknowledge alert.
```

FIGURE 6.3   Use of a real-time alert engine to notify providers of critical results and potential interventions. (Courtesy of Brigham and Women's Hospital, Boston, Massachusetts.)

experience: recent vital signs, nursing observations, current orders, etc. The information needs to be grouped based on the "affiliation" of the patient (e.g., inpatient-based teams or the attending physician responsible for his or her inpatient care).

Information must be delivered to providers even when the patient is not in front of them in a hospital bed or in a clinic examination room. Many systems now provide clinicians with a repository of results or messages, often in a manner similar to an electronic mail "in box," so that findings from recent encounters can have appropriate follow-up (Figure 6.3). Some types of result posting require more immediate feedback (e.g., in inpatient settings). Some hospitals have embarked on projects to page providers upon receipt of any results.[15] In other instances, notification only occurs if critical results are posted to a database or a serious interaction between a medication order and a laboratory result is indentified (see Figure 6.4).[16]

```
   CAREPLAN ,FIVE         MR#: 19114784  6E        6106A    05/07/09  0741
      70792197   F    12/12/1932  Age:   76  MGM    ACTIVE PLAN OF CARE
                               ─ Plan of Care ─

        Description                                           Disp  Date
 □  P: NUTRITION IMBALANCE R/T: > BODY REQUIREMENTS                 05/07
 □    EO: PT WILL MAINTAIN OPTIMAL WEIGHT & NUTRITIONAL BALANCE     05/07
 □     I: COLLABORATE W/DIETITIAN FOR PT NUTRITION/EDUCATION        05/07

 □  P: EDU: PT/FAMILY EDUCATION/DC PLANNING                         05/07
 □    EO: LEARNER DESCRIBES DX/POC/DC PLANNING BASED ON PT NEEDS    05/07
 □     I: CM: HOME CARE ARRANGED (SEE CASEMGR NOTE)                 05/07




                  Disp  Web  Rev/Res  Enter   Add   D/C
 Main  Reslts CarePln  Note  Ref  POC  Notes  EO/Int  Int  Print
```

FIGURE 6.4   Interdisciplinary plan of care. (Courtesy of Siemens Medical Solutions, Malvern, Pennsylvania.)

6.4 DOCUMENTATION

The most tangible outcome from a provider visit is often the "note" that emerges from an encounter with a clinician summarizing the findings and assessments of that visit in an office or inpatient setting. This note is used as a record for legal purposes to describe the care provided and as a communication tool to other providers. The note can further serve as justification for the charges attendant to that encounter; insurers often ask for copies of this documentation as part of auditing to ensure that the services were in fact rendered.

To meet these multiple needs of clinical documentation, many EHR products have tools to facilitate documentation so that discrete clinical elements can be easily incorporated into notes. Aside from the narrative of the patient history, the note can often contain prior diagnostic information about a patient, current medications, current laboratory values, and recent vital signs. Because elements of prior family and social history elements are incorporated, the "complexity" of the note increases and providers are given credit for this extra work.

The challenge in this form of note production is that it is only efficient for the provider at the time of the encounter if all of the discrete medical information referenced has been already placed in the record in other settings. One of the key frustrations in adopting this kind of template note creation is the large up-front investment of time. Additionally, because much of the structured note entry requires the provider to describe the narrative using preformed phrases, clinicians might feel this is too great a reduction in free expression and that they are unable to convey the richness of the patient's story.[17]

As identified previously, nursing documentation in the inpatient arena often has a different flavor and presents an ongoing record of interventions to produce desired outcomes in a given patient. Nurses often outline a series of problems based on initial admission assessments; from this, they derive interventions to ameliorate the problems. This structured approach allows for a common space where multiple ancillary providers (e.g., respiratory therapists, case managers, social workers, dietitians, physical therapists) can interact in one area to describe the ongoing progress of the patient toward health (see Figure 6.4).

Evidence is increasing that use of these standardized plans is seen as an improvement in the approach to the hospitalized patient by nurses.[18] The EHR allows for linking of a given problem to several note entries, regardless of the originating discipline. As mentioned previously, the greatest challenge to this approach to documentation is that physician providers, who are required to provide discreet "notes" of their individual encounters for fiscal reasons, have found it challenging to adopt this more problem–intervention-based methodology for charting.

6.5 ORDER MANAGEMENT

To understand the EHR's real impact on order management, one first has to start with an understanding of the previous paper-based order management system. In the case of medication ordering, the usual (and still prevalent) method is for a physician or other licensed provider to write a prescription or medication order manually on paper, for that paper to be delivered in some manner to a pharmacy, and then for the pharmacist to transcribe or enter that order into another system—either an ambulatory pharmacy benefits application or an inpatient pharmacy system, depending on the setting. Only at that point does the order get the benefit of being checked against structured terminology, existing drug databases, and advanced clinical decision support systems. Of course, at the moment of initial writing or in transcription, there is chance for error on the part of the physician, pharmacist, or other pharmacy personnel.

The fate is similar for orders for other services and may even have additional chances for error. In the case of an inpatient laboratory order, the provider might write a paper order for a given test that then has to be written onto a manual requisition for the laboratory. This is then handed to a laboratory technician, who transcribes the order into the laboratory system. This model has two transcription moments rather than one, doubling the chance of a transcription error. It was in this context that the first attempts were made to streamline the order process—first in inpatient and then in ambulatory settings.

The first hospital systems focused on clerk order entry to mitigate the rewriting of physician orders onto requisitions. Having clerks enter these orders into a core system that would then route the orders to the appropriate system (e.g., laboratory, radiology, or, less often, pharmacy) offered at least a controlled vocabulary of tests and procedures from which to choose and increased likelihood of appropriate test ordering. However, this still created one degree of difference from the moment of clinical thought (on the part of the physician) to the moment of execution (order entry).

To eliminate this step, a small number of hospitals in the late 1980s developed computerized provider order entry (CPOE) modules to attach to their EMR products. The goal of CPOE was to provide structure to physician orders and to allow for the rendering of decision support tools at the moment of ordering.[19] The development of CPOE in subsequent years has been to extend the tools of standardized vocabulary and passive and active decision support to all areas of physician ordering (described more fully in Chapter 15). This development came to include areas such as medication ordering and study ordering, and it went on to cover specific directions to nurses for bedside care.

The other key distinguishing feature of CPOE is that it ensures that the correct provider received the result. A case in point is the management of a laboratory order. A physician can enter a laboratory order and when the result is sent back to the core EHR, it will route automatically to the results folder of the ordering physician, ensuring both continuity of care and appropriate follow-up of results. Of course, as with any system implementation, unintended consequences can occur.

In current hospital environments—when many physicians are working in shift mode rather than in long blocks of being "on call"—the ordering physician may no longer be the appropriate person to contact. For example, Dr. A may order a certain drug level and when the result comes back in toxic range, a page will be sent to Dr. A to alert the provider; however, if Dr. A's shift is over, the system has to shift seamlessly to the covering provider rather than the ordering provider. In this instance, many CPOE systems still provide benefit by tagging orders with the provider who entered it as well as the patient about whom the result is posted. A separate table can list current "covering physicians" (see Figure 6.5).

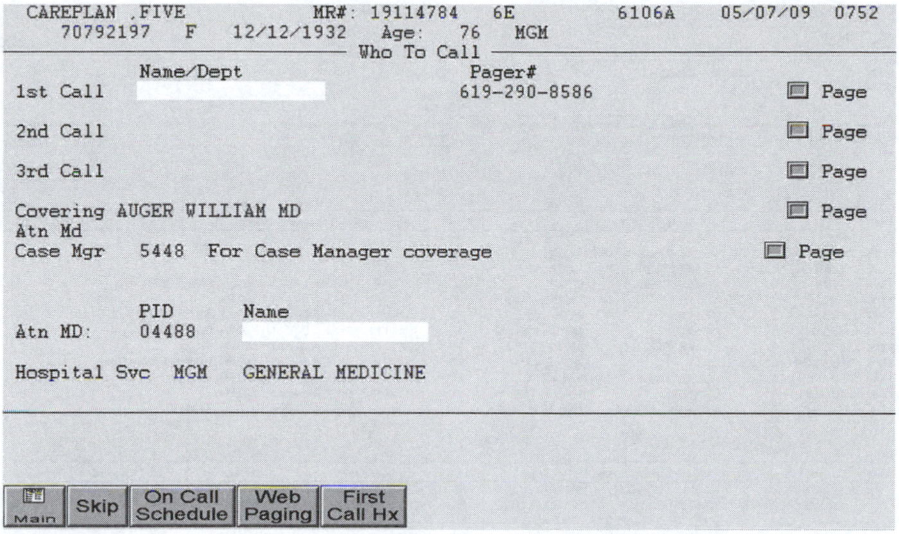

```
CAREPLAN ,FIVE          MR#: 19114784   6E         6106A    05/07/09  0752
    70792197   F   12/12/1932  Age:   76  MGM
                           ─ Who To Call ─
           Name/Dept                      Pager#
1st Call   ┌──────────────┐               619-290-8586            ☐ Page
           └──────────────┘

2nd Call                                                          ☐ Page

3rd Call                                                          ☐ Page

Covering AUGER WILLIAM MD                                         ☐ Page
Atn Md
Case Mgr   5448  For Case Manager coverage                       ☐ Page


           PID        Name
Atn MD:    04488      ┌──────────────┐
                      └──────────────┘

Hospital Svc  MGM    GENERAL MEDICINE
```

┌──────┬──────┬──────────┬─────────┬──────────┐
│ ▦   │ Skip │ On Call │ Web   │ First   │
│ Main │   │ Schedule │ Paging │ Call Hx │
└──────┴──────┴──────────┴─────────┴──────────┘

FIGURE 6.5 How to contact the covering physician. (Courtesy of Siemens Medical Solutions, Malvern, Pennsylvania.)

6.6 CPOE AND MEDICATION ORDERING IN THE EHR

Medication ordering is a unique subset of orders that has received the most attention because most sources identify medication management as the most error-prone element of both inpatient and ambulatory care (see Chapter 10 for more details).[4,20] Because it is the best described and represents the largest penetration of CPOE in the inpatient environment, we will focus on the complete order life cycle for medications in an inpatient EHR.

6.6.1 Inpatient Drug Orders

The physician initiates the order, entering a series of discrete elements such as dose, frequency, route, priority (e.g., "stat"), start and stop times, and other key instructions picked from a structured drug dictionary as referenced earlier in this chapter. Sent across an interface, the order is received in a work queue by a pharmacist, who might see additional alerts to help him or her determine the clinical safety and validity of an order. In most pharmacy systems linked to the EHR, the pharmacist can see the entire medication list as well as pertinent laboratory values (to assist in pharmacokinetic or other dosage calculations).

Within their own pharmacy systems, pharmacists are often in a position to modify the order according to local policy or on drug availability, provided that it is an approved therapeutic equivalent to the original order. In a small percentage of hospitals, the order goes forward to a medication administration system after validation. The system then provides the bedside nurses with an active work list of medications to be administered—the medication administration record (MAR; see Figure 6.6). In combination with a bar coding system to identify the patient and the medication to be administered, the administration is checked for correctness of time, patient, route, dose, and frequency one more time (see also Chapter 8). This step of checking the administration one more time before drug administration reduces the risk of preventable adverse drug events.[21]

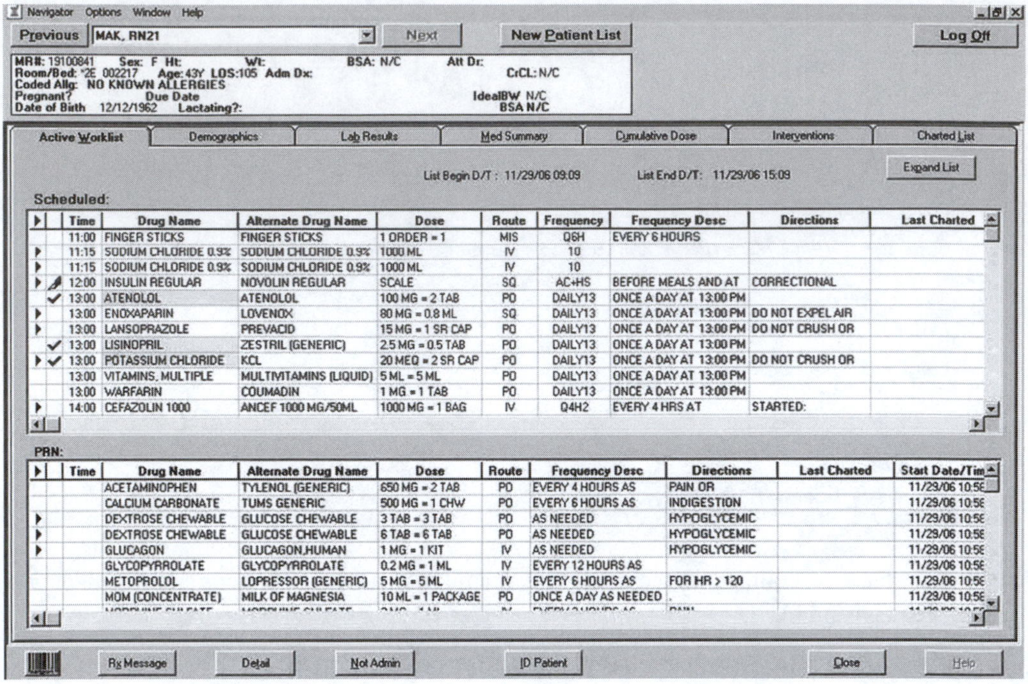

FIGURE 6.6   Nursing view of all medications in an active work list that enables bar coded administration for safety. (Courtesy of Siemens Medical Solutions, Malvern, Pennsylvania.)

Finally, in some systems, the loop is fully closed as documentation of the medication administration is passed to another view for the physician provider so that he or she can see the precise time at which the medication was given. In this manner, the EHR leverages the use of structured terminology, decision support at the time of ordering, and validation to improve the safety of the order and the quality of the documentation.

6.6.2 Outpatient Drug Orders

The life cycle of a medication order in an ambulatory EHR replicates many of the physician ordering steps, but often has a different fate when it is sent to an outpatient pharmacy. The ambulatory physician must consider not only the medication dosage information, but also whether a given medication is covered by the patient's insurance. It is much less likely that the ambulatory medication order will be transmitted directly to a pharmacy, so a paper or facsimile intermediary is produced.

Increasingly, the U.S. government is interested in promoting electronic prescribing with physician order entry as well as direct links to pharmacies. A federal rule mandating financial incentives for the use of this tool for patients who receive federally funded healthcare has been published.[22] The EHR plays a role because it has the capacity to maintain the patient's financial information in a structured format as well as the list of preferred pharmacies to promote electronic transmission.

6.6.3 Decision Support

The precise manner in which electronic records support safer drug prescribing is relatively well studied and focuses on information provided to the ordering physician at the time of order placement. For example, such support can improve the rate at which patients achieve a therapeutic drug level. It is also clear that giving recommendations in real time leads to providers adjusting a drug's dosage. However, it has been less easy to demonstrate that these supports help keep a patient in therapeutic range or that such a range can be reached more rapidly using decision support.[23]

Most current leaders in clinical informatics recognize that clinical decision support still has shortcomings and call for increased effort to improve the user interface to make suggestions more intuitive and to target alerts so that providers are not offered so many alerts that the meaning of any individual warning is reduced[24] (see Chapter 15).

A growing body of literature examines the unintended consequences of provider order entry, such as new work for clinicians, unfavorable work flows, new errors through the presentation of inconsistent data, and impaired communication between co-workers because of excess reliance on computerized communication.[25]

6.7 SECONDARY USE OF THE EHR TO PROMOTE QUALITY

Now that the functional components of the electronic record have been defined, it is vital to understand the way in which the record has come to be used to drive forward initiatives in overall quality of care. Much as in clinical decision support associated with medication ordering, the judicious use of alerts and reminders is intended to prompt the physician to take action (to order) in a manner consistent with established, evidence-based guidelines (see Chapter 13).

These schema have been well studied in the ambulatory arena where key quality indicators exist in the domain of cancer screening, glycemic management for persons with diabetes, and lipid management for those with coronary artery disease. A common scenario is that of a provider seeing a patient with identified coronary artery disease on the problem list. The EHR "knows" the most recent lipid levels as well as whether the patient is taking a medication to lower lipid levels. If the patient is not currently on an antilipemic agent or does not have a serum lipid level consistent with national guidelines, an alert will display (see Figure 6.7). Another screen will prompt the prescriber to order appropriate lipid-lowering therapy (see Figure 6.8). Although previous indirect evidence indicated the benefits of alerting, more direct randomized, controlled trial evidence now shows improved lipid management through these systems.[26]

On a more global level, the record provides an aggregation of all care delivered and can target specific providers or practices that appear to be delivering care less consistent with national guidelines. Many institutions embark on the creation of a "clinical data warehouse" to capture all clinical data derived from an electronic record for later analysis (see Chapter 16). The benefit of analyzing data after they have been moved out of the EHR is that the data can be reviewed without affecting the performance of the database used to

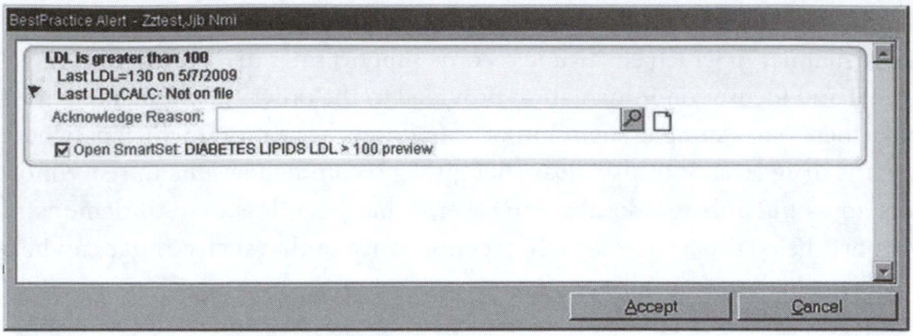

FIGURE 6.7   Provider being alerted that lipid-lowering medication is indicated. (Courtesy of Epic Systems, Madison, Wisconsin.)

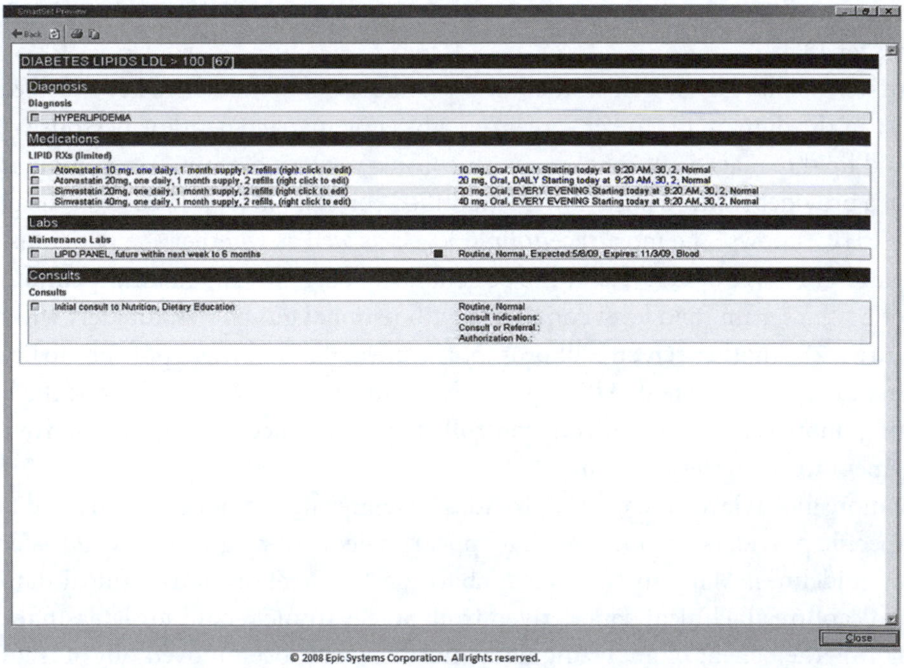

FIGURE 6.8   Provider is being assisted in complying with lipid-lowering guidelines. (Courtesy of Epic Systems, Madison, Wisconsin.)

care for patients. This also allows for merging clinical data with financial, genomic, or other data not in the EHR.

Although systematic reviews have identified that electronic records have improved the consistency, accuracy, and completeness of the chart, it is less clear what the impact has been on patient outcomes.[27] In fact, in one review of care delivered at ambulatory sites in 2003 and 2004 in which 18% of visits were associated with an EHR, no statistical difference was found in 14 of 17 nationally established quality indicators between visits with and without EHR use.[28] Thus, there is evidence on a local level of improved care quality, but this has yet to translate across a large group of heterogeneous users.

6.8 SUMMARY

The EHR is now the established communication tool for healthcare delivery in the twenty-first century. Electronic health records are based on structured healthcare terminology, accommodate the practices of documentation and provider–provider and patient–provider communication, and provide for order management. The new frontier of patient health records and the intersection with clinic- or hospital-based electronic records is as yet to be fully explored, but will likely define the next generation of clinical informatics. To date, the EHR has a proven track record of improving medication safety and record accuracy, legibility, and completeness. However, the jury is still out on the large majority of patient-level outcomes that were to be improved by the advent of electronic records and electronic orders. The full potential of these novel tools in improving outcomes remains to be proved in multisite, randomized, controlled trials.

REFERENCES

1. Dexheimer, J. W., Talbot, T. R., Sanders, D. L., Rosenbloom, S. T., and Aronsky, D. Prompting clinicians about preventive care measures: A systematic review of randomized controlled trials. *Journal of the American Medical Informatics Association* 2008. 15:311–320.
2. Durieux, P., Trinquart, L., Colombet, I., et al. Computerized advice on drug dosage to improve prescribing practice. *Cochrane Database Systems Review* 2008. 3:CD002894.
3. Tang, P. C., and McDonald, C. J. Electronic health records. In *Biomedical informatics,* 3rd ed., ed. E. H. Shortliffe and J. J. Cimino. New York: Springer, 2006.
4. Bates, D. W. Using information technology in hospitals to reduce rates of medication errors. *British Medical Journal* 2000. 320:788–791.
5. Bates, D. W., Evans, R. S., Murff, H., Stetson, P. D., Pizziferri, L., and Hripcsak, G. Detecting adverse events using information technology. *Journal of the American Medical Informatics Association* 2003. 10:115–128.
6. Hammond, E. W., and Cimino, J. J. Standards in biomedical informatics. In *Biomedical informatics,* 3rd ed., ed. E. H. Shortliffe and J. J. Cimino. New York: Springer, 2006.
7. http://www.ama-assn.org/ama/pub/physician-resources/solutions-managing-your-practice/coding-billing-insurance/cpt/about-cpt.shtml
8. Huff, S. M., Rocha, R. A., McDonald, C. J., et al. Development of the Logical Observation Identifier Names and Codes (LOINC) vocabulary. *Journal of the American Medical Informatics Association* 1998. 5:276–292.
9. Henry, S. B., Holzemer, W. L., Randell, C., Hsieh, S. F., and Miller, T. J. Comparison of nursing interventions classification and current procedural terminology codes for categorizing nursing activities. *Image: Journal of Nursing Scholarship* 1997. 29:133–138.

10. McDonald, C.J., Overhage, J. M., Tierney, W. M., et al. The Regenstrief medical record system: A quarter century experience. *International Journal of Medical Informatics* 1999. 54:225–253.
11. Jha, A. K., DesRoches, C. M., Campbell, E. G., et al. Use of electronic health records in U.S. hospitals. *New England Journal of Medicine* 2009. 360:1628–1238.
12. Kaushal, R., Jha, A. K., Franz, C., et al. Return on investment for a computerized physician order entry system. *Journal of the American Medical Informatics Association* 2006. 13:261–266.
13. Branstetter, B. F., IV. Basics of imaging informatics. Part I. *Radiology* 2007. 243:656–667.
14. Bellon, E., Feron, M., Peeters, K., et al. Eliminating the paper medical archive by bulk document scanning of historic folders and implementing revised work flows for scanning new documents. *Studies in Health Technology Informatics* 2008. 141:121–129.
15. Poon, E. G., Kuperman, G. J., Fiskio, J., and Bates, D. W. Real-time notification of laboratory data requested by users through alphanumeric pagers. *Journal of the American Medical Informatics Association* 2002. 9:217–922.
16. Jha, A. K., Kuperman, G. J., Teich, J. M., et al. Identifying adverse drug events: Development of a computer-based monitor and comparison with chart review and stimulated voluntary report. *Journal of the American Medical Informatics Association* 1998. 5:305–314.
17. Linder, J. A., Schnipper, J. L., Tsurikova, R., Melnikas, A. J., Volk, L. A., and Middleton, B. Barriers to electronic health record use during patient visits. *AMIA Annual Symposium Proceedings* 2006. 499–503.
18. Dahm, M. F., and Wadensten, B. Nurses' experiences of and opinions about using standardized care plans in electronic health records—A questionnaire study. *Journal of Clinical Nursing* 2008. 17:2137–2145.
19. Kuperman, G. J. et al. *HELP: A dynamic hospital information system.* New York: Springer-Verlag, 1991.
20. Kaushal, R., Goldmann, D. A., Keohane, C. A., et al. Adverse drug events in pediatric outpatients. *Ambulatory Pediatrics* 2007. 7:383–389.
21. Morriss, F. H., Jr., Abramowitz, P. W., Nelson, S. P., et al. Effectiveness of a bar code medication administration system in reducing preventable adverse drug events in a neonatal intensive care unit: A prospective cohort study. *Journal of Pediatrics* 2009. 154:363–368.
22. Electronic Code of Federal Regulations. Title 42: Public Health. http://ecfr.gpoaccess.gov/cgi/t/text/text-idx?c=ecfr&sid=dcea4421d09168675bd2e194df2f0176&rgn=div5&view=text&node=42:3.0.1.1.10&idno=42.
23. Durieux, P., Trinquart, L., Colombet, I., et al. Computerized advice on drug dosage to improve prescribing practice. *Cochrane Database Systems Review* 2008. 16:CD002894.
24. Sittig, D. F., Wright, A., Osheroff, J. A., et al. Grand challenges in clinical decision support. *Journal of Biomedical Informatics* 2008. 41:387–392.
25. Campbell, E. M., Sittig, D. F., Ash, J. S., Guappone, K. P., and Dykstra, R. H. Types of unintended consequences related to computerized provider order entry. *Journal of the American Medical Informatics Association* 2006. 13:547–556.
26. van Wyk, J. T., van Wijk, M. A., Sturkenboom, M. C., Mosseveld, M., Moorman, P. W., and van der Lei, J. Electronic alerts versus on-demand decision support to improve dyslipidemia treatment: A cluster randomized controlled trial. *Circulation* 2008. 117:371–378.
27. Häyrinen, K., Saranto, K., and Nykänen, P. Definition, structure, content, use and impacts of electronic health records: A review of the research literature. *International Journal of Medical Informatics* 2008. 77:291–304.
28. Linder, J. A., Ma, J., Bates, D. W., Middleton, B., and Stafford, R. S. Electronic health record use and the quality of ambulatory care in the United States. *Archives of Internal Medicine* 2007. 167:1400–1405.

CHAPTER 7

Pharmacy Information Systems

Daniel T. Boggie, Jennifer J. Howard, and Armen I. Simonian

CONTENTS

7.1 INTRODUCTION

Pharmacy information system support for automation has become vital to optimize the safety and efficiency of the medication use process. Today's practice of pharmacy incorporates many technologies to assist the pharmacist in delivering care. Increased efficiency reduces pharmacist time in preparing and dispensing medications and allows increased pharmacist time for clinical activities. The adoption of these technologies also allows deployment of advanced safety measures.

The pharmacy system incorporates patient-specific clinical data to support review of medication appropriateness, perform real-time inventory management, and interact fully with other systems, including computerized provider order entry (CPOE), bar code medication administration (BCMA), automated dispensing systems, billing systems, and electronic health records (EHR). Pharmacy information systems also exist in the ambulatory setting and serve to help pharmacists receive, process, fill, and dispense prescription orders; track inventory; and bill insurance payers for prescriptions.

Determining information and technology needs is challenging in complex environments such as healthcare. Information solutions can be adopted on an enterprise or systemwide basis, or stand-alone solutions may be introduced separately and interfaced together. Stand-alone systems run independently to support a specific function and can continue to function when other systems experience downtime; however, they must interface with other systems to communicate information properly. Integrated systems do not require separate interfaces to communicate information to other areas, but may cease to function when the system experiences downtime. Integrated systems may be less customizable or offer fewer specialized features than stand-alone systems.[1]

Whether fully integrated or stand-alone systems are implemented, they should meet the specific needs of the organization. There are functions specific to hospital settings and to ambulatory settings. Institutions may require support only for inpatient or outpatient functions; however, some integrated institutions may require a comprehensive system that supports both care settings. Pharmacy work flow, handling of orders, and dispensing differ between the two settings. A pharmacy information system must be able to handle functions in both environments or individually as required.

In this chapter, we will examine the components and function of a hospital pharmacy information system, an outpatient pharmacy information system, and the use of automation in the medication use process in both settings.

7.2 HOSPITAL PHARMACY INFORMATION SYSTEMS

Medication use within an inpatient setting requires coordinated and timely communication among numerous healthcare disciplines. Ideally, all pertinent data are available on demand and accessible by pharmacists and other healthcare personnel. Pharmacy information systems must interface directly with other components of a clinical information system to communicate data to and from prescribers, pharmacy, laboratory, and nursing. Integrated communication greatly enhances efficiency of pharmaceutical care delivery and increases patient safety by optimizing information necessary to ensure safe and appropriate medication use within a facility.

Medication use in a hospital setting may be described simply in several steps: ordering, verification, dispensing, administering, and monitoring. Information is used and generated at each step in this continuum, and pharmacy information systems must be able to provide the information needed as well as to communicate and document the information generated.

Traditionally, pharmacists received handwritten physician orders on specific pharmacy order forms that were physically delivered to the pharmacy for processing. Technological advances allowed the capture of these forms to be transmitted directly to pharmacy via facsimile or other digital image capture and transmission software. However, the limitations of handwritten orders remained. Today, pharmacists may receive prescriber orders electronically directly from a CPOE system, which allows prescribers to place pharmacy and other orders for care electronically. CPOE systems can integrate with a pharmacy information system in a variety of ways that include four basic options:

1. A *fully integrated CPOE and pharmacy information system* will have the same basic structure and necessary interfaces so that communication between the two systems

is seamless. Stand-alone CPOE or pharmacy information systems require an interface for electronic communication between the two systems.

2. A *bidirectional interface* between CPOE and the pharmacy system allows orders to be transmitted electronically to an existing pharmacy information system and for orders to be entered directly into the pharmacy information system and transmitted back through CPOE to populate other clinical systems, such as automated dispensing equipment or medication administration systems. This functionality allows pharmacists to take verbal or telephone orders from prescribers in urgent clinical situations, as well as order corrections, and to enter the orders in lieu of a prescriber who is offsite or unavailable to enter the order.

3. A *unidirectional interface* would allow orders to be transmitted only from the CPOE system to the pharmacy information system, eliminating the ability of pharmacists to enter or take action on prescriber orders.

4. *Not using an interface* between CPOE and the pharmacy information system would require the printing of orders from CPOE in the pharmacy. Pharmacy personnel would then enter the orders into the pharmacy information system manually, introducing the possibility of transcription errors.[1]

Once orders are received by the pharmacy, the order verification process begins. During this process, information from data sources provides the pharmacist with patient demographic, allergy, laboratory, and other pertinent clinical data. Potential problems with prescribed therapy including drug–drug, drug–food, and drug–disease interactions are screened and corrected or acknowledged by the pharmacist. Verified orders may then be sent to automated dispensing or dose preparation equipment and to BCMA systems to prepare for medication administration by nursing staff (see Chapter 8). Finally, appropriate billing or charge data are generated and processed either at the time of administration or dispensing. After administration, monitoring information from the bedside and laboratory will be evaluated by the pharmacist to ensure the safety and efficacy of therapy.

Several components of an integrated pharmacy information system must have bidirectional communication to share and record healthcare data generated during a hospital stay (see Figure 7.1). Central to all hospital activities is the patient. A patient file will contain all primary patient data, including name, sex, age, allergies, height, weight, attending physician, primary and secondary diagnoses, and insurance information.

Admission, discharge, and transfer (ADT) information is essential for the hospital setting and is provided by an ADT system. Patients often change beds, rooms, wards, or units, so accurate and up-to-date patient location information is critical. A prescriber file contains all authorized prescribers within the hospital and data including license numbers, prescribing privileges, medical specialty, and pager number. An actively maintained list of medications available for prescribing within the hospital is found in a drug file. A drug file may have many associated subfiles that contain vital drug information, including doses, warnings, available routes of administration, criteria for use, formulary status, and additional dispensing comments.

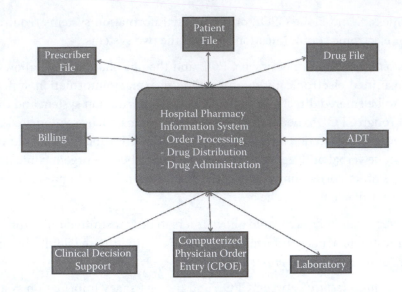

FIGURE 7.1   Hospital pharmacy information system architecture.

A pharmacy information system must interface with the laboratory to provide pharmacists with pertinent data to support assessment of patient status and appropriateness and outcomes of therapy—for example, the patient's renal and hepatic function to assess when monitoring drug therapy. Additionally, a laboratory interface can provide culture and sensitivity information to help guide antimicrobial therapy selection and therapeutic drug levels necessary for accurately dosing potentially toxic drugs. Clinical decision support (CDS) systems use information from each of these sources—patient, drug, laboratory, ADT, and prescriber—to provide clinicians with integrated data to help guide or support clinical decision making. Interface of CDS with a pharmacy information system is vital to help guide medication use (see Chapter 15).

An integrated pharmacy information system can also interface with drug distribution automation to ensure the timely and accurate availability of medications to be selected and administered by nurses. The pharmacy information system may also interface with medication administration programs, such as BCMA, or an electronic medication administration record (eMAR). Providing accurate and timely information from the pharmacy information system to administration programs better equips nursing to meet the "five rights" of medication administration effectively: right drug, right dose, right route, right patient, and right time (see Chapter 10).

Finally, an integrated pharmacy information system may interface with coding and billing systems to ensure accurate documentation and billing for pharmacy resource consumption. This type of interface can ensure that patients and insurance providers are charged only for the medications used during an inpatient stay.

7.3 OUTPATIENT PHARMACY INFORMATION SYSTEMS

Outpatient pharmacy information systems are complex computer systems used by community and ambulatory care clinic pharmacies to process, fill, and dispense prescriptions.

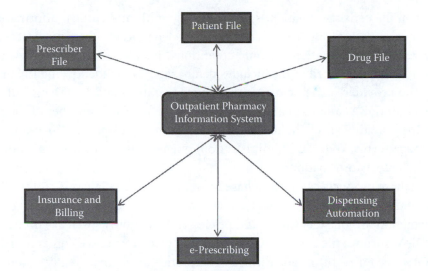

FIGURE 7.2   Outpatient pharmacy information system architecture.

Outpatient pharmacy information systems can be integrated into hospital information systems or other points of care to allow transmission of prescriptions from the prescriber directly to the pharmacy. The majority of computer systems in community pharmacies are stand-alone systems designed to support the function of an individual pharmacy or chain of pharmacies. Many outpatient pharmacy information systems are commercially available and selection of the most appropriate system can depend on whether the pharmacy is a chain or independent pharmacy.

Over 50% of pharmacy chains in the United States use pharmacy software developed by their corporate office or by the PDX Pharmacy System.[2] Chain pharmacy systems have developed the ability to transfer and share prescription information electronically across the chain rather than relying on telephone transfers. Enabling patients to fill their prescriptions at any location enhances convenience for patients. Electronic transfer of prescription information markedly streamlines this process for pharmacy chains. Smaller, independent pharmacies typically purchase a stand-alone system (e.g., QS/1 or McKesson Pharmacy Systems).

The main components of an outpatient pharmacy information system are similar to those of the hospital systems, but functions specific to the outpatient pharmacy setting, such as insurance billing for prescription claims, are also supported (Figure 7.2). Generally, these computer systems consist of the following components:

- patient file;
- prescriber file;
- drug file;
- insurance information file and billing interface;
- e-prescribing interface; and
- dispensing automation interface.

The patient file consists of patient-specific demographic and clinical information, such as patient name, date of birth, sex, address and telephone number, medication allergy and drug reaction information, chronic conditions, insurance information, and a prescription profile. The prescriber file includes demographic data identifying the prescriber, including the provider name, address, and phone numbers and the Drug Enforcement Administration (DEA) and/or National Provider Identifier (NPI) number of the provider. The drug file contains all of the drugs that the pharmacy dispenses, including product names, dosage forms, routes of administration, strength, unit of measure, and procurement and inventory information.

Drug database products can be purchased from companies such as First DataBank and Medispan and incorporated into the pharmacy information system to provide descriptive drug information, unique identifiers, and pricing information, as well as clinical decision support. The insurance file contains a list of all supported insurance companies as well as a billing interface for real-time adjudication of prescription claims. Finally, automated dispensing equipment interfaces may send completed prescription data to a robotic dispenser for filling.

Outpatient pharmacy information systems may be similar in structure to hospital systems, but the process used to verify and fill prescriptions is quite different from start to finish. In the hospital setting, medication orders are entered by the provider into the computer system via CPOE. In contrast, prescriptions in the outpatient setting are most often handwritten by a provider and carried by the patient to the pharmacy. New prescriptions can also be faxed to the pharmacy or called in by the provider's office.

Handwritten, faxed, or telephoned prescriptions must be transcribed into the pharmacy information system by the pharmacy staff once the prescription is received. The pharmacy staff is required to verify all pertinent patient information, provider information, and insurance and benefit information and select the appropriate drug.

E-prescribing is the process by which a provider can electronically send an accurate, error-free prescription directly to the pharmacy information system from the point of care. E-prescribing greatly enhances patient safety and in the future will be the standard for transmission of new prescriptions from the prescriber to the pharmacy. These prescriptions are automatically transcribed into the pharmacy information system and displayed for pharmacist processing.

The prescription is ready for processing by the pharmacist once it is entered into the pharmacy information system. It is during this step that allergy, drug–drug interaction, and drug–disease interaction checking automatically occurs. With CDS and order-checking processes, the pharmacy information system can determine whether the dose of the drug is appropriate for the patient's weight, age, or renal function. The pharmacy information system can also screen for potential problems with prescribed therapy, including duplicate therapies, drug–drug, drug–food, and drug–disease interactions, and notify the pharmacist.

Unlike hospital pharmacy information systems, real-time online insurance claim processing and adjudication occur during outpatient prescription processing (see Figure 7.3). If the patient has prescription drug coverage on file, the pharmacy information system

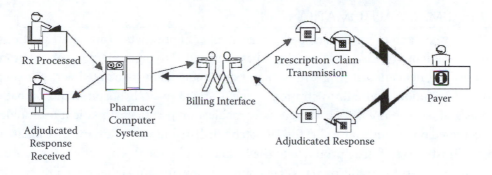

FIGURE 7.3   Online insurance claim processing.

will send the prescription claim information to the insurance company's central computer system through a billing interface. The prescription is then checked against stored insurance coverage data to determine whether it can be approved for payment. The pharmacist is immediately notified if the prescription is not covered by the patient's insurance, if it is too early to refill the prescription, if a prior authorization (PA) is needed, or if a drug utilization review (DUR) is required before the fill can be processed.

Once the prescription has been processed by the pharmacist, a label can be printed and placed on the prescription. Alternatively, the prescription information can be sent to an automated drug-dispensing machine, which then fills and labels the vial. Automated dispensing machines are discussed later in this chapter.

All prescriptions must be checked by a pharmacist before they are dispensed to the patient. This can be a manual process, or it can be facilitated by the pharmacy information system. Many pharmacy information systems print a bar code on the prescription label that is scanned at the point of pharmacist checking. The bar code scan brings up a digital image of the drug for the pharmacist to aid in verification of the correct medication.

Pharmacies are required to keep documentation of patient counseling according to applicable state and federal laws. Most outpatient pharmacy information systems now have electronic signature pads to capture patient signatures for documentation of patient counseling. An electronic signature can be stored within the pharmacy information system for as long as required.

In the outpatient pharmacy setting, the prescription is dispensed directly to the patient or his representative. This is most often accomplished by the pharmacy staff handing the prescription to the patient at the pharmacy window. Many pharmacy information systems have point of sale (POS) software incorporated into their systems that help manage inventory. Some systems have the ability to track the status of a prescription (e.g., whether the prescription is currently being filled or is sitting on the shelf). There is also automation for dispensing to patients when the pharmacy is closed. These pharmacy dispensing "kiosks" are discussed in the next section of this chapter.

Once a prescription has been filled by a pharmacy, refill prescriptions can be entered into the pharmacy information system manually, received by an automated telephone system, or processed through a Web-based program that interfaces with the pharmacy information system.

7.4 PHARMACY AUTOMATION

Technology companies have developed a large number of products to automate portions of the medication use process. Software logic has been added to machinery to create automation tools that help alleviate the large amount of manual labor associated with the preparation, distribution, and administration of medications in both hospital and outpatient settings and, at the same time, promote safer medication practices. Inpatient technologies have centered on the unit-dose cart fill, IV batch fill, distribution to nursing care areas, and safe administration of drug products to the hospitalized patient. Outpatient devices have helped to automate the traditional tasks of counting, pouring, and labeling of prescriptions and the distribution of the finished product to the patient.

7.4.1 Inpatient Pharmacy Automation

Before the concept of unit-dose dispensing became a standard, bulk packages of medications were kept in locked medication rooms in nursing care areas. Often referred to as ward stock, individual doses were poured by nurses from the bulk bottle into cups from which these doses were administered to individual patients. With the advent of the 24-hour cart exchange process, a patient's daily doses were added to a cassette and delivered to the care area at a given time each day. Medications were dispensed directly by the pharmacy and the pharmacy consequently had a need for individually packaged medications.

Early automation included packaging machines with labeling and database software to create and document production lots of unit-dose tablets and capsules from the manufacturer's bulk bottles. The next step in automation was to interface the packaging machine to the pharmacy information system, enabling the machine to package and group all of the patient's 24-hour supply of medications. These were then placed in the unit-dose cassette, ready for delivery after verification by a pharmacist.

7.4.1.1 Automated Dispensing Cabinets

An alternate concept of drug distribution eventually emerged, and these new machines returned to the old idea of ward stock. In this form of automation, the ward stock was in unit-dose packaging and placed in discreet locations within the drawers of a locked cabinet. The cabinet drawer locks were controlled by a computer and electromechanical system that was able to open drawers, indicate the location of the medication within the drawer, sense the return of the drawer to a closed position, and record the entire dispensing transaction including a date and time stamp and identification of the practitioner (usually a nurse), patient, and medication dispensed. This technology is now known as the automated dispensing cabinet (ADC).

Competing vendors have taken different approaches to the dispensing methodology of ADCs. One refinement has been to restrict access to a drawer full of medications by exposing only one drug at a time to the practitioner. This feature has been accomplished in various ways, such as individually covered, locked locations (or pockets) that open only for the selected drug when the drawer opens. Another method is the use of a rotating mechanism or sequential drawer that opens to only one medication or dose at a time. Finally, some

ADCs have mechanisms similar to those on vending machines, which allow the device to gather an individual, selected medication and drop it into a receptacle from which the practitioner can retrieve the dose. This method eliminates the need to pick the dose from a location in the drawer and avoids errors associated with the selection process.

Medication dispensing safety has been enhanced with ADCs by the introduction of gatekeeper logic, which maintains a database of active orders for each patient and will only allow dispensing of medications for which valid orders exist. Though the ADC may contain the top 300 medications used in a particular area, the practitioner is only able to select medications for removal from a list of active orders for a given patient.

With this gatekeeper logic activated, the ADC may provide a bypass mechanism to allow for dispensing in emergent situations or other special circumstances. The software may allow an exception or override list of medications that are always available for dispensing to any patient. In these cases, the practitioner may select a medication for which a valid order has not yet been recorded for the patient within the ADC database. Standard reports are generated to help the pharmacist follow up on all of these special dispensing events and verify that a valid order for the medication was recorded in the patient's medical record.

Additional medication safety enhancements include rules and alerts. Some ADC software can alert the practitioner in cases where the patient has a recorded allergy to the drug selected for dispensing. Rules and alerts can also be developed to warn the practitioner of possible adverse outcomes and ask for a response to document the justification for continuing with the dispensing event after a warning has been displayed.

Specialized ADCs have been created for areas such as anesthesiology, where the ADC replaces a medication tray or cart in the operating room. The anesthesiologist is afforded quick access to medications and simple methods for documenting administration during the procedure. Software for these cabinets is geared toward multiple dispensing events on a single patient over an extended period of time in contrast to the typical "one patient–one drug" removal of medications by nurses using traditional hospital ADCs.

7.4.1.2 Smart Pumps

When IV infusion pumps were first introduced, their main function was to provide infusion of piggyback, syringe, or large-volume medication at a specified rate. Eventually, software and interfaces were added to these devices to create what are now known as intelligent or smart pumps. These newer devices are able to store a database of standard IV preparations, allowing the nurse to select from a list of predefined items when ready to administer an IV solution to a patient. Preparations listed in the database can be assigned default rates with CDS parameters that can warn of rates too low or too high.

With both ADCs and smart pumps, detailed information is saved with each transaction. In addition to the recording of date, time, patient, practitioner, medication, and other basic data, the triggering of alerts and warnings, responses to override or bypass questions, and breaches of dosage limits may be recorded. These data can be used to evaluate medication usage, monitor for drug diversion, investigate potential and actual adverse events, and help identify targets for quality improvement initiatives. One of the major criteria for the evaluation of these systems for selection and implementation is the standard reporting functionality and usefulness of report generation tools.

7.4.1.3 Robots

Pharmacy robots take automation one step further with the addition of kinetic components that mimic human activity. Early attempts were made to produce and market multifunctional robots that handled and integrated many of the central steps of the medication use process, but these products were not successful. Hospitals now purchase and deploy robots that address specific functions—mainly preparing, dispensing, and distributing.

Dispensing robots can accept medications in the form of individual, unit-dosed, barcoded product and stock their internal storage locations using robotic arms integrated with bar code readers. The robot can then be used to select, label, and supply first doses to be dispensed, assemble 24-hour cart-fill medications for individual patients, provide refill medications for ADCs based on par levels, or perform a combination of these activities.

Preparation robots can mimic sophisticated human activities such as the production of patient-specific IV solutions or the repetitive, bulk production of stock catheter flushes, antibiotics, and other standard syringe medications. These machines also use robotic arms and bar code readers, but in certain cases add cameras for the documentation of steps and components. An IV robot can accept a vial of medication, sterilize the rubber stopper, reconstitute the medication, withdraw and add the medication to an IV solution bag, and label the final product. At each step, components can be verified with the use of bar code technology, and cameras can take digital photographs of additive and solution labels and document the robot's product selection, sterile technique, and other steps of the process.

Distribution robots add the component of mobility to the machine. These robots can replace the delivery person, pneumatic tube system, or other methods of delivering medications. Medications can be loaded into the robot, which is able to navigate the hospital corridors and elevators and deliver the medications to the nursing care areas. Nursing personnel can load returned medications and solutions into the robot, and these returns can be transported back to the pharmacy. The distribution robot is able to call elevators using infrared technology and use sensors and internally stored maps to navigate to the desired destination. Sensors can also detect human or other obstructions, helping the robot to avoid collisions.

Bar code scanning—a relatively simple technology—has become a key component of the functionality and integration of pharmacy automation. Ensuring that the right drug goes to the right location within an ADC can be verified by the scanning of both product and drawer-location bar codes during loading and refilling activities. Bar code technology is an essential component of IV preparation and robotic systems.

The convergence of all of these technologies with the use of bar code verification occurs at the point of care with the use of BCMA at the bedside. When a hospital has implemented the preceding pharmacy automation systems with corresponding interfaces, the nurse at the bedside has the opportunity to perform "five rights" verification with every unit-dose medication given and every IV infusion administered to the patient.

7.4.2 Outpatient Pharmacy Automation

Community pharmacies have been using automation for decades to improve the efficiency and accuracy of drug filling and dispensing. The automatic tablet counter was among the earliest automation vehicles used in pharmacies. At present, automation is available commercially for all aspects of the prescription filling and dispensing process. Devices that assist with medication selection and counting, pouring, and labeling of prescriptions are in use. In addition, automation that dispenses finished prescriptions to the patient when the pharmacy is closed has been developed. This section will discuss the two major types of outpatient automation: automated drug-dispensing machines and automated kiosks.

7.4.2.1 Automated Drug-Dispensing Machines

Automated drug-dispensing machines (ADDMs) interface with the pharmacy information system and have the ability to fill, label, and deliver prescriptions. Once the prescription has been processed by the pharmacist, the complete prescription information can be sent electronically to an ADDM. Most ADDMs hold only tablets and capsules. ScriptPro's SP 200 contains 200 dispensing cells for tablets and capsules of all sizes (see Figure 7.4). After filling the prescription, the SP 200 prints and applies the prescription and auxiliary labels. The uncapped vial is delivered to the pharmacist for final checking using digital drug image verification.

Baker cells are another commonly used ADDM that can also be interfaced with the pharmacy information system. When the prescription information is sent from the pharmacy information system, tablets or capsules are automatically counted into a chute. The pharmacist then opens the chute to release the medication into a prescription vial.

7.4.2.2 Automated Kiosks

Automated kiosks are ATM-style machines that deliver drugs to patients. These machines do not fill the prescription; they only hold the medication that has been filled by the

FIGURE 7.4   ScriptPro's SP 200.

FIGURE 7.5   ScriptCenter.

pharmacist until the patient can pick it up. Kiosks give patients the convenience of picking up and even paying for their prescriptions with a credit card at any time, even when the pharmacy is closed. Asteres' ScriptCenter is an example of a commercially available kiosk (see Figure 7.5).

Initially, patients are required to enroll with the pharmacy to have their prescriptions placed in ScriptCenter. Then the patient can order refill prescriptions as usual. When the patient presents at the ScriptCenter kiosk, he or she is prompted to enter a user ID and PIN or to scan a fingertip and enter a PIN. Then, the patient is required to review and select the prescriptions to pick up. The final step is to sign for, acknowledge, and pay for the prescriptions. Once payment is approved, ScriptCenter delivers the prescriptions to the patient. A phone number for after-hours pharmacist consultation is posted on the ScriptCenter as well as printed on the customer receipt.

7.5 SUMMARY

Pharmacy information systems and automation fulfill important roles in supporting an efficient and safe medication use process in both hospital and outpatient settings. Integration between all sectors of care and between all settings where care is given will be a great challenge as technology and information use in healthcare continues to evolve. Data standardization will become important as integration of healthcare information systems progresses.

Information and automation can improve the efficiency of pharmacy operations by reducing the time that pharmacists spend in manual aspects of dispensing functions while not sacrificing oversight and quality control. An efficient medication use process creates an opportunity for pharmacists to provide more cognitive, clinical services. Technology can also dramatically improve the accuracy and safety of pharmaceutical care delivery, leading to desirable patient outcomes and reduction in potentially dangerous errors.

REFERENCES

1. Chaffee, B. W., and Bonasso, J. 2004. Strategies for pharmacy integration and pharmacy information system interfaces. Part I: History and pharmacy integration options. *American Journal of Health-System Pharmacy* 61:502–506.
2. Skrepnek, G. H., Armstrong, E. P., Malone, D. C., and Abarca, J. 2006. Workload and availability of technology in metropolitan community pharmacies. *Journal of the American Pharmacists Association* 46(2):154–160.

CHAPTER 8

Bedside Bar Coding Technology and Implementation

Ashley J. Dalton

CONTENTS

8.1   INTRODUCTION

The Institute of Medicine estimates that medication errors contribute to thousands of deaths each year.[1] Although mistakes in healthcare can occur by many different means, one study estimated that 34% of medication errors occur and could be prevented at the point of administration.[2] In efforts to decrease preventable errors, many healthcare institutions are beginning to use technology to support the medication administration process. Processes that employ bar code medication administration (BCMA) are emerging as technological

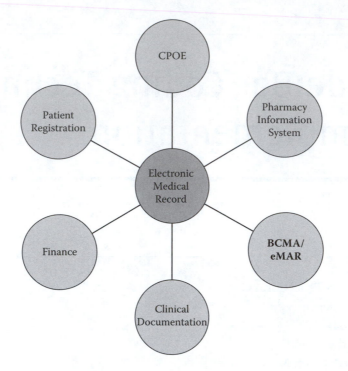

FIGURE 8.1   BCMA as a component of an electronic health record.

improvements to traditional medication administration procedures. In 2007, a study from a large medical center (>400 beds) detailed a 54% reduction in medication administration errors through BCMA implementation.[3]

8.1.1 BCMA Procedural Overview

BCMA comprises both hardware and software designed to enhance the accuracy of medication delivery to a patient. Hardware components can include mobile computers at each bedside and printing systems for placing bar codes on patient identification bands, as well as on medications. Scanning hardware is also used in BCMA. These devices range from simple linear scanners, capable of reading only linear bar codes, to very complex digital imagers that are able to read a variety of dimensional bar code symbology.

BCMA hardware works in conjunction with BCMA software that is usually housed on a server. Incoming information is processed on the server and returned to a bedside computer that is tied to a communications network (usually wireless). As illustrated in Figure 8.1, many BCMA applications are also interfaced to other institutional medical record databases, such as the pharmacy information system, computerized provider order entry system (CPOE), and the finance or billing system.

BCMA operates using bar-coded patient identification bands and bar-coded medications to be documented in the patient's medication administration record (MAR) or electronic MAR (eMAR). An eMAR is an important component of BCMA because it holds all of the documented medication administrations completed using BCMA. See Figure 8.2 for an example of an eMAR.

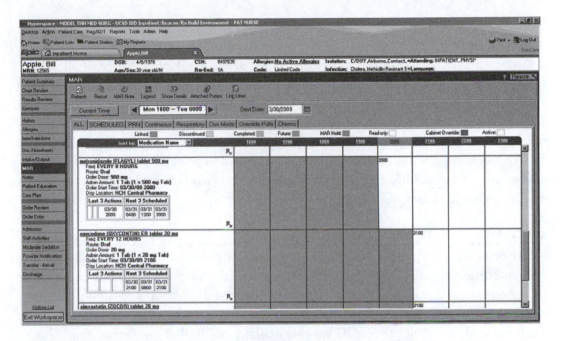

FIGURE 8.2   Sample eMAR. (EPIC Systems Corporation. Reproduced with permission.)

Some institutions also use BCMA to document administration of nonmedication entities, such as blood products and pumped breastmilk, to achieve best-practice standards set forth by accreditation agencies. A BCMA system patient record usually contains a complete medication profile that includes all of the medications, scheduled and as needed, that have been ordered by a provider for administration during the hospital stay, as well as the eMAR. A BCMA patient record may also contain specified laboratory results, patient demographic information, or other features that are system specific.

A provider intending to administer a medication logs into the system and then selects the patient's name from an admissions listing or scans the bar code listed on the patient's identification band to activate the patient's record. Then the provider selects the name of the medication to be given from a list and scans the bar code on the medication or simply scans the medication, forcing the system to search the patient medication profile for the drug. Once the medication is correctly identified by the system, the user is prompted to scan the patient's identification bar code if he or she did not previously do so.

In addition, many systems have the capability of creating best-practice alerts for the user. These alerts could include specific messages related to the medication being administered or even require a second user to serve as a witness or double-check for the administration of a medication (e.g., cancer chemotherapy). BCMA technology helps nurses quickly identify the "five rights" of medication administration: patient, drug, dose, route, and time of administration. Once all five rights of medication administration are correctly identified, the user is invited to administer the medication and appropriately document the occurrence in the eMAR. Some institutions may even have their BCMA system programmed to bill the patient upon documenting administration of a medication. This ensures that only doses that have been administered to a patient are billed.

8.1.2 Prevalence

Although BCMA demonstrates great potential in medication administration error reduction and over 20 companies have developed or are developing BCMA applications, the technology is still widely underused.[4] According to a national survey of pharmacy practice in hospital settings performed by the American Society of Health-Systems Pharmacists in 2008, only 25% of hospitals use BCMA.[5] However, this represents a substantial increase from only 1.5% of hospitals using BCMA in 2002.[6] This increase in BCMA use is expected to continue to rise as more studies reveal the importance of medication use safety technology. It is therefore important that pharmacists understand bar coding technology and applications.

8.2 BAR CODE BASICS

A bar code is a group of printed bars and spaces designed to be scanned and read into computer memory; they contain information about the labeled object. Bar codes come in various sizes, shapes, and formats. The bar code type or symbol is commonly referred to as symbology. Currently, more than 20 different types of bar code symbologies exist. Each bar code symbology presents specific benefits and limitations. Some symbologies allow only for numeric content; others allow for alphanumeric content. Content storage capacity also varies by symbology. For example, the most widely used bar code symbology, EAN-13, can encode 13 numerical characters and is used for the marketing of retail goods worldwide.

Two-dimensional bar codes have a large capacity for data storage. For example, Maxicode can encode 93 data characters and was designed for rapid automated scanning of packages traveling at speeds of 500 feet per minute on a conveyor belt. Even more impressive is the two-dimensional bar code PDF-417, which can encode over 1,000 characters in one small bar code. (See Figure 8.3 for additional examples of bar code symbologies.) Due to the drastic variance in bar code symbologies, agencies such as the Uniform Code Council have set standards for global commerce of retail items. Currently, no standardization model exists for bar codes placed on medications; until 2004, there was no bar coding requirement for drug manufacturers.

In February 2004, the U.S. Food and Drug Administration (FDA) published a final ruling that requires certain human drug and biological products to have a linear bar code on their labels.[7] The bar code, at a minimum, is to contain the *National Drug Code* (NDC) number. Manufacturers with drugs approved before the final ruling were given 2 years to comply with the request. The intent of the FDA ruling is to help prevent medication errors through the use of BCMA technology by healthcare organizations. The FDA estimated that the ruling could reduce the number of medication errors that occur by 500,000 over the next 20 years.[8]

Although the FDA ruling is an important first step in aiding the use of BCMA technology, it is vague in some areas and does not contain the specific requirements that would allow hospitals to use drug manufacturer bar codes for BCMA better. Three areas of the ruling that need more specificity are bar code content, over-the-counter items, and the unit

BAR CODE	TYPE
	SSCC-18
	CodaBar
	EAN-128
	EAN-13
	EAN-14
	RSS Expanded
	RSS Expanded Stacked
	SCC-14
	UCC-128
	UPC-A
	USS CODE 128

FIGURE 8.3   Examples of bar code symbology.

dosing of products. Unit dosing refers to the act of reducing medications packaged in bulk containers to packages intended for single-patient use.

8.3 BUMPS IN THE ROAD

Implementation of BCMA in a hospital is a herculean and expensive endeavor. Direct capital costs to an organization can start between $1 million and $5 million; adding in training, infrastructure, and other hidden costs, the number can approach $10 million or more.[4] The decision to implement BCMA at an institution is not only a financial commitment to purchasing hardware and software, but also an institutional commitment to create a new culture focused on patient safety. BCMA implementation can be one of the largest changes to affect nursing processes and nursing patient care in an organization. In addition to nursing, the pharmacy department is also greatly affected by a decision to implement BCMA. In the BCMA environment, the pharmacy must be ready to change existing procedures and implement new ones to support a successful launch of BCMA fully.

A careful analysis of each step of the medication management process and how it relates to BCMA should be performed prior to implementation. Steps in the medication management process include prescribing, dispensing, administering, monitoring, educating, and outcomes. Such an analysis can reveal potential direct and indirect changes to pharmacy processes and internal quality assurance initiatives.

8.4 BCMA MEDICATION MANAGEMENT FOCUS

8.4.1 Prescribing and Pharmacy System Order Entry

The success or failure of a bar code scan at the bedside often relies on the way in which an order was input into the hospital or pharmacy information system. Depending on how an order is entered, the BCMA system will anticipate what types of data the bar code for that order should include. This could include an NDC or designated drug code that has been created for a customized drug record in the pharmacy information system. For example, a customized drug record may be needed for noncommercial drug items that are compounded by the pharmacy. These items do not have an NDC number and must be assigned a drug code to complete the drug record in the pharmacy information system.

Another source of bar code data that the BCMA system may anticipate is one that has been forced as a recognizable entry in the pharmacy information system or computerized BCMA library. A forced entry can occur when the bar code data for a manufacturer bar code is unrecognizable to the system. Many BCMA systems are equipped with mapping tables that allow the user to map various bar code data streams to a specified drug record.

Another source of bar code data may come from patient-specific bar codes generated by a pharmacy information system. Patient-specific bar codes are primarily used for combination products, such as compounded intravenous solutions containing more than one ingredient, and may contain both patient identification and medication order data. BCMA implementation requires that pharmacists have knowledge of how specific medications and preparations are correctly entered to ensure successful verification at the bedside. In addition, invalid medication order entries into the BCMA system that must be reentered correctly can cause cluttering of the eMAR.

Order entry errors or duplicate medication orders can cause unintended entries on the eMAR that require an additional action by the user. For example, an order for a one-time dose of acetaminophen 650 mg orally is sent to the pharmacy. The pharmacist enters the order into the pharmacy information system for two 325 mg acetaminophen tablets to be given now. However, after processing the order, the pharmacist realizes that the patient has a nasogastric tube and the liquid formulation would be a more appropriate form of medication for the patient. The pharmacist then reenters the order using the liquid formulation, resulting in two entries for acetaminophen on the eMAR. In this scenario, the pharmacist and the nurse must communicate to ensure that two doses are not administered, and the nurse must also perform the separate action of documenting the invalid entry as not administered to the patient.

Invalid entries on the eMAR can cause BCMA user bar code overrides, phone calls to the pharmacy, and a loss of faith in the BCMA system. All elements of order entry in the pharmacy information system and how they affect the eMAR must be carefully reviewed prior to BCMA implementation. It is imperative that pharmacists have an understanding of the pharmacy information system used by their organization and how it interfaces with BCMA.

Another critical factor in BCMA that can have an impact on both nursing and pharmacy is the administration time of medications. One of the five rights of drug administration includes giving the medication at the right time. All scheduled medications in the BCMA system will have a specific time for administration. In the environment of the hospital, the time that the medication is specified to be administered may not be the time at which it can be administered. For example, one tablet of aspirin (325 mg) is due to be administered to a patient at 0900, but the nurse is unable to administer the medication because the patient is getting an x-ray and is not in the patient care unit.

Although most healthcare institutions have standard times of medication administration, it is up to the institution to determine how to handle the retiming of medications. These decisions may force the organization to determine definitions for what may be considered "late" or "early" administration. Most BCMA systems allow some latitude on the administration time of orders, prompting users with a warning if the administration is outside the defined time parameters. Some systems employ indicators, such as color changes on the patient eMAR, signifying that a medication order is late and should be administered in a timely fashion.

Depending on the pharmacy information system and its interaction with the BCMA system, the pharmacy may be responsible for the retiming of medication orders. Prior to implementation, a structured methodology regarding the appropriateness of medication order retiming should be implemented by the institution. This could include instructions on proper notification of the time change to the pharmacy or a way for nursing to "catch up" to the correct administration time without permanently changing the entire time schedule.

8.4.2 Dispensing

Analysis of the dispensing of medications needs to encompass every activity involved with a medication from a "door-to-patient" perspective. This analysis can include drug procurement, repackaging, and relabeling. One of the biggest challenges to the pharmacy is deciding how to achieve the goal of placing a bar code on every medication that is intended for direct patient administration.

A few methods can be used to achieve this goal. Medications can be procured from many different vendors. The pharmacy procurement focus changes after an organization begins to implement BCMA. It no longer includes only cost and what is inside the package, but now incorporates the packaging itself. Most wholesalers have been astutely monitoring the development of BCMA and, in anticipation, have developed categories within their purchasing catalog databases that include an indication of products available in unit-dose form. Although this does not give the user an indication whether the product has a readable bar code or even a bar code at all, it is a helpful resource.

Many manufactured products are available in unit-dose form. When looking at manufacturer unit-dosed products, the pharmacy must take into consideration the cost difference between bulk and unit-dosed medications, the increase in costs that the department is willing to accept, and the need for pharmacists and technicians to perform activities other than the preparation of unit-dosed bar coded (UDBC) products. In addition, the concept of packaging quality assurance must come into consideration. Pharmaceutical manufacturers and third-party repackaging entities must abide by standards set forth by regulatory agencies for good manufacturing practices. It is very difficult for a hospital pharmacy to have the same quality assurance practices as a manufacturer when medications are packaged.

In preparation for BCMA implementation, most pharmacies examine their shelves to estimate the amount of labor for manual preparation of UDBC products from products packaged in bulk. Many pharmacies are pleasantly surprised to find that 60–80% of the products on their shelves are available in unit-dose form. However, in a survey conducted by the Institute of Safe Medication Practices, 6% of respondents reported having difficulty obtaining unit-dose medications from manufacturers that had previously provided unit-dose products.[9] These manufacturers had changed their packaging back to bulk form, forcing hospitals to unit-dose bar code the product themselves. In the same survey, 75% of respondents reported having difficulty with the packaging of received manufacturer UDBC products. The 2004 FDA bar code ruling does not require manufacturers to unit-dose products.

In addition, if a manufacturer does choose to provide UDBC products, the ruling does not place strict requirements on bar code symbology or bar code content. Manufacturers use many different types of symbology and many of them place data other than the NDC in the content of the bar code. Additional data often include product lot and expiration date, but they can also be internal data relevant only to the manufacturer.

The receiving of UDBC products from outside sources forces many pharmacies to introduce new quality assurance procedures into their product receiving processes. Typically, once a product order is received, the products can be placed on the pharmacy storage shelves or directly into automated dispensing cabinets. In an environment with BCMA, each product must be checked for a readable bar code at the unit-of-use level. This means that at least one of every UDBC product received must be scanned and compared to the hospital BCMA software to ensure that the bar code is correctly recognized. This quality assurance step can cause major pharmacy work-flow changes.

The diagram in Figure 8.4 is one representation of a typical pharmacy receiving process. The boxes shaded in light gray represent pharmacy procedures without BCMA. Each box shaded in dark gray represents a new step in the product receiving process.

Instead of or in addition to purchasing manufacturer UDBC products, institutions can choose to repackage and unit-dose bar code every medication internally. According to a survey by the American Society of Health-Systems Pharmacists, 93% of hospitals repackage some oral medications into a unit-dose form.[5] Methods for unit-dose bar coding oral medications range from almost completely manual to highly automated. Manual processes usually involve a pharmacy technician placing tablets, capsules, or specified quantities of pharmaceutical liquids into unit-dose cups or bags. Labels or stickers are then generated using UDBC software. At a minimum, the labels usually contain the medication name, strength, dosage form, lot number, beyond-use date, and a usable bar code.

Highly automated methods can include the use of high-speed packagers that can produce UDBC medications in large quantities in a very short period of time. Some packaging machines can accommodate large canisters for storage of 100–500 different oral solids and have packaging speeds of 20–60 doses per minute. Many high-speed packagers have the capability to store medications in canisters and use the machine's bar coding technology; this allows the user to scan the bulk product and a bar code on the canister to verify correct placement of the oral solid. In either style of repackaging, the data encoded in the bar code

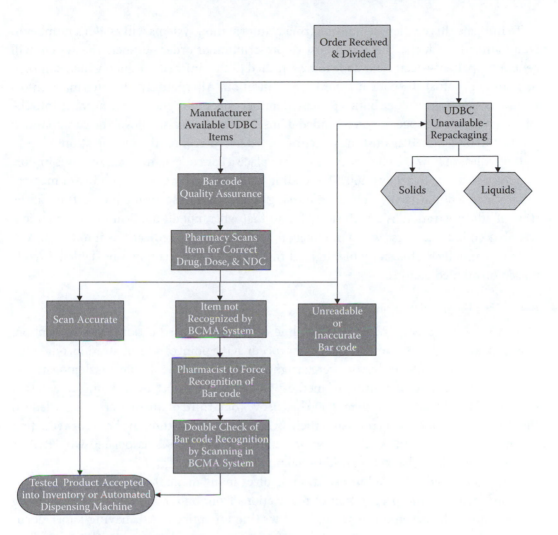

FIGURE 8.4 Pharmacy product receiving for BCMA.

can be configured based upon UDBC software's capabilities and the institution's choice of BCMA software.

Alternatively, all UDBC responsibilities could be outsourced to a repackaging company. Yet another option for achieving 100% bar coding for all medications is for an institution to purchase as much manufacturer UDBC product as possible, repackaging or relabeling products only when absolutely necessary. Each of the methods described previously has advantages and disadvantages that must be matched to the institutional needs. Most pharmacies choose a combination of the three methods.

An additional challenge presented with BCMA occurs when the pharmacy compounds a product with multiple ingredients. This is most common with intravenous (IV) admixtures and formulations compounded for infants and children. The challenge is to decide what data should be included on the bar code of the compounded product. Many admixtures contain several components in varying amounts (e.g., parenteral nutrition) that could not all be represented in a single bar code.

To mitigate this challenge, many pharmacy information systems will assign a number to compounded products. This is similar to a prescription or order number. The system will generate a patient-specific bar code to be attached to the finished product, which can then be scanned against the patient record for verification. Alternatively, the pharmacy information system may encode both a patient identifier and an assigned order number into the labeling for patient-specific compounded items. This method also allows the compounded product to be scanned against the patient's record for correct product verification.

Regardless of how the pharmacy chooses to place a bar code on medications, it is important for it to strive to achieve 100% bar coding on all medications. The number of medications bar coded may have an effect on nursing compliance in scanning medications at the time of administration. A Dutch study found that when not all medications for a patient were bar coded, the nurse was more likely to enter all of the medications into the BCMA system manually.[10] This study highlighted the importance of having a readable bar code placed on all medications.

8.4.3 Drug Administration

Nurses are often recognized as the last line of defense in medication error prevention. However, although 38% of medication errors occur at the point of administration, one study found that only 2% of these errors were caught.[2] BCMA implementation has demonstrated success in increasing prevention of medication errors. In a post-BCMA implementation study in a large healthcare system (>500 beds), over one-third of nurse users acknowledged that they had avoided an error using their BCMA system.[11] Although the error reduction potential has been acknowledged, the procedures and processes associated with BCMA impose tremendous changes on traditional nursing practice.

Traditional medication administration involves many manual processes. Manual process can include nurse preparation of medications from bulk containers (e.g., preparing doses from bulk suspension bottles), hand-writing an entire MAR, having paper documents at hand prior to medication administration, and manual documentation of medication administration on the MAR. Prior to BCMA implementation, current nursing work flow should be documented in detail and compared with the anticipated BCMA integrated work flow. Nursing leaders can then focus on the largest areas of change and work toward solutions and staff education.

Performing a careful work-flow analysis is essential in addressing nursing perceptions of the changes supported by BCMA and anticipating BCMA procedural "work-arounds." One of the most often cited oppositions to BCMA is that it will increase the amount of time that nurses spend on medication administration activities. However, a time and motion study performed at a 735-bed tertiary care hospital found that, prior to BCMA implementation, nurses spent approximately 26.9% of their time performing duties associated with medication administration. This compared with 24.9% of their time spent on medication administration after BCMA implementation—not a statistically significant change.[12]

Actual or perceived inefficiencies can lead to nurses' deviations from the established BCMA procedures. These often include overriding bar codes on both mediations and patient ID bands, scanning medication bar codes away from the bedside, printing multiple patient

ID bands to bypass scanning one physically attached to the patient, saving bar codes from the packaging of frequently used medications, and manually keying in data located in a bar code rather than physically scanning the bar code. A bar code "override" refers to the act of the user intentionally not scanning the bar code on the medication or patient ID and selecting the option of bypassing that safety check. Most BCMA systems allow overriding of bar codes, but require the user to input a reason for the occurrence (e.g., hardware failure, ripped bar code). Implementation teams strive for procedural excellence; however, care must be taken to prevent bar code overrides and work-arounds and to develop ways to combat them.

Despite the potential for drastically reducing medication administration errors, improper use of a BCMA system negates any potential benefits. It is imperative that nurses support and are fully satisfied with the BCMA system selected by an organization.

8.4.4 Monitoring

One of the most important steps in the medication use process is monitoring the effects of the medication. In other words, what was the outcome of administering the medication? BCMA systems often have configurable monitoring components to facilitate or even prompt monitoring. These system settings can include forced or passive documentation.

Examples of forced documentation include a requirement to obtain a patient's blood sugar measurement prior to insulin administration or checking blood pressure prior to administering an antihypertensive. An example of passive documentation is capturing the volume of parenteral fluid given to a patient through the administration of multiple intravenous medications for inclusion in total patient fluid intake. Often, this recorded information will be translated for other parts of the electronic health record at the time of documentation for integrated or interfaced systems. This is beneficial for nurses because it can save time by eliminating redundant documentation.

Monitoring components of a BCMA system can play an important role in a patient's treatment plan. For example, in treating a patient having difficulties with pain management, most institutions require the nurse to ask the patient to rate the pain before giving a pain medication. In some BCMA systems, the nurse is able to document a patient's pain score and, in some cases, respiratory rate at the same time as administering the analgesic. As part of monitoring the medication's effects on the patient, a second pain score and respiratory rate may be taken after a predetermined amount of time has passed since the drug was given. This effectively enables providers to assess the safety and efficacy of a particular medication.

BCMA and an eMAR are particularly useful to providers when medications that require pharmacokinetic monitoring are administered. At the moment a medication is administered, the exact time of administration can be captured by the BCMA system and recorded on the eMAR. Instead of manual documentation or searching through multiple paper MARs, the information needed to relate drug administration times with the subsequent serum drug concentrations can be accessed from a computer.

8.4.5 Education

In many healthcare institutions, nurses are the first line of information for patients regarding their medications. In addition, many regulating bodies require that patients be told what medication they are being given and the indication for its use. This requires that the nurse have a basic knowledge of drug information or have an easily accessible drug information source. For this reason, many BCMA systems are equipped with links to basic drug information monographs or have links to drug information databases to which the institution subscribes. Some systems take patient education a step further by providing printable drug information handouts written at a typical patient comprehension level.

8.4.6 Outcomes

The most important outcome of BCMA systems is preventing medication errors at the time of drug administration. The FDA has predicted that BCMA systems can reduce medication errors by as much as 50%, but the agency has not elaborated on the specific types of errors that may be prevented. One study categorized errors prevented by BCMA by collecting the error logs generated from BCMA systems at six different community hospitals and having a review panel examine the logs.

The review panel found that the majority of medication errors detected by the BCMA system were benign (i.e., early or late administration of medications). Fewer than 10% of the errors prevented were expected to cause moderate or severe harm to the patient. Although this percentage seemed low, when that number was extrapolated to the 18 million doses administered with the BCMA systems in this study, it still effectively prevented 17,000 medication errors predicted to cause moderate or severe harm to a patient, thus affirming the benefits of BCMA.[13]

BCMA systems have the ability to capture an enormous amount of reportable data. Nursing performance measures can be readily available and might include timeliness of medication administration or undocumented medication doses scheduled for administration. Measures specifically related to BCMA can also be generated and may include reports on overridden drug bar codes, overridden patient identification bands, overridden system warnings, or overridden cosignatures.

8.5 IMPLEMENTATION STRATEGIES

Many different healthcare groups have recommendations for implementing BCMA and tips for using it. Of particular note are the American Society of Health-Systems Pharmacists Foundation and the Institute for Safe Medication Practices (ISMP).[14,15] Common themes emerge from these recommendations. In order to conduct a successful implementation of BCMA, an organization must create a multidisciplinary BCMA team, conduct a formal readiness assessment for the entire organization, and begin a shift in culture toward patient safety and awareness.

8.5.1 Creating a BCMA Team

BCMA spans many disciplines. In order to have a successful implementation, an organization is advised to include all of the vested parties in planning. Although the level of participation may vary, each representative may have a function that will need to incorporate BCMA. The total size of the team should be relatively small (6–10 people). This group should be considered the project lead team and each team member may have assignments requiring him or her to form action groups outside the project lead team in order to achieve the goals. The following disciplines are suggested for inclusion on a BCMA implementation team; however, each institution should create a team specific for its need that takes into account all of the projected users and their work areas.

Because of the tremendous change to daily work flow that BCMA presents to nursing, a nurse is a logical selection as a task force chair and project leader. It is imperative that nursing provide strong leadership and commitment to the implementation of BCMA. When nurses are satisfied with their point-of-care technology, they may be less likely to develop work-arounds. If the end user does not back the product, it will not be successful.

Another natural fit for a BCMA project lead and cochair on the BCMA task force is a pharmacist. BCMA affects all parts of pharmaceutical management and the pharmacy has to be ready to incorporate bar coding into each aspect flawlessly. Pharmacy is the backbone that supports BCMA. Nursing success and ease of system use are partially dependent on the pharmacy. If all medications do not have a readable bar code, the primary functionality of a BCMA system is lost.

An informatics leader is another natural selection to sit on the project leadership team. Financial and technical support for both the hardware and software of BCMA is essential.

Ancillary members of the project team may include people from departments such as hospital quality improvement, respiratory therapy, nursing education, medicine, purchasing, and materials management. Depending on the services they offer and their configuration, some institutions might choose not to implement BCMA in departments that care for a mix of inpatients and outpatients, such as radiology or a postanesthesia care unit. Personnel from these areas may then not need to be part of the team.

8.5.2 Conduct a Formal Readiness Assessment

The purpose of a readiness assessment is to examine the organization for assets and work flows essential to BCMA and determine at what state of readiness each currently functions. Items may be categorized as fully ready, partially ready, or not ready. The concept behind the assessment is to take each item that is fully ready and develop a plan to maintain the current status. Items that are only partially ready or not ready will need a plan to move to a fully ready state.

The ISMP has developed a readiness assessment that encompasses nine essential elements of BCMA.[15] The assessment presents each element as a detailed checklist. Some of the items on each checklist are listed as prerequisites, and others are listed as facilitators. Prerequisites are items that must be in place prior to BCMA implementation; facilitators are not required, but would most likely make implementation easier. See Table 8.1 for a listing of the nine essential elements of BCMA.

8.5.3 Begin a Culture Shift toward Patient Safety and Awareness

A workforce culture can be defined by the shared basic assumptions about an organization's values, beliefs, and behaviors that have been taught to personnel.[16] The Agency for Healthcare Research and Quality (AHRQ) defines patient safety as freedom from accidental or preventable injuries caused by healthcare.[17] Therefore, a culture of patient safety

TABLE 8.1   ISMP Elements BCMA Implementation Success

ISMP nine elements related to successful BCMA implementation	Example
Drug labeling, packaging, and nomenclature	Determine the most frequently used medications requiring a bar code
Drug standardization, storage, and distribution	Ensure that all patient care areas are using unit dose medications
Environmental factors	Ensure information systems are secure with access control
Patient information	All patient wristbands have a bar code that includes a unique patient identifier
Drug information	The BCMA system is able to provide maximum dosage alerts for high-alert drugs, such as chemotherapy
Communication of drug orders and other drug information	Standardized medication administration times have been implemented
Staff competency and education	A formal training and competency program has been established
Patient education	Educational resources have been dedicated to helping patients understand BCMA and what it means for their care
Quality processes and risk management	A business case has been developed for BCMA and agreed upon by senior organizational leaders

can be defined as shared values, beliefs, and behaviors that place a priority on preventing healthcare-related injuries. A culture of patient safety is paramount for the successful implementation of BCMA. Users of technology that promote patient safety may be more inclined to use the technology properly if the organization culture is striving for excellence in patient safety.

To define and measure a culture of patient safety more clearly, the AHRQ began developing patient safety culture surveys in 2004. The survey elicits information from 12 areas considered essential in a culture of patient safety[17]:

- communication openness

- feedback and communication about error

- frequency of events reported

- handoffs and transitions

- management support for patient safety

- nonpunitive response to error

- organizational learning

- continuous improvement

- overall perceptions of patient safety

- staffing, supervisor expectations, and actions promoting patient safety

- teamwork across units and

- teamwork within units.

Organizations can use the data reported by the AHRQ to benchmark themselves against others and learn where improvements are needed within themselves to optimize a culture of patient safety.

Organization-wide understanding and awareness of medication errors can facilitate a culture that anticipates and demands implementation of error-preventing technologies. Institutions proactively participating in healthcare culture assessments, such as the survey conducted by the AHRQ, have the knowledge to begin developing an action plan for improvement and facilitating implementation of patient safety technologies such as BCMA.

8.6 SUMMARY

Thousands of deaths or injuries occur every year as a result of medication errors. Some of these errors can be attributed to errors in the medication administration process. BCMA is a rapidly emerging technology that can reduce errors in this process. The implementation of BCMA has a large impact on the work flow of both nurses and pharmacists. Each step in medication management should be explored for the impact imposed by BCMA. Implementation is best facilitated by creating a multidisciplinary team to lead the project, conducting an organization-wide readiness assessment, and fostering a culture of patient safety and awareness.

REFERENCES

1. Institute of Medicine. *To err is human: Building a safer health system.* Washington, D.C.: National Academy Press, 1999.
2. Bates, D. W., Cullen, D. J., Laird, N. et al. Incidence of adverse drug events and potential adverse drug events: Implications for prevention. *JAMA* 1995. 274:29–34.
3. Paoletti, R. D., Suess, T. M., Lesko, M. G., et al. Using bar-code technology and medication observation methodology for safer medication administration. *American Journal of Health-System Pharmacy* 2007. 64:536–543.
4. Cummings, J., Bush, P., Smith, D., et al. Bar-coding medication administration overview and consensus recommendations. *American Journal of Health-System Pharmacy* 2005. 62:2626–2629.
5. Pedersen, C. A., Schneider, P. I., and Scheckelhoff, D. J. ASHP National Survey of Pharmacy Practice in Hospital Settings: Dispensing and Administration—2008. *American Journal of Health-System Pharmacy* 2009. 66:926–946.
6. Pedersen, C. A., Schneider, P. I., and Scheckelhoff, D. J. ASHP National Survey of Pharmacy Practice in Hospital Settings: Dispensing and Administration—2002. *American Journal of Health-System Pharmacy* 2003. 60:52–68.
7. Department of Health and Human Services Food and Drug Administration. Bar code label requirements for human drug products and biological products; final rule. *Federal Register* 69 (38): February 26, 2004.
8. Department of Health and Human Services Food and Drug Administration. FDA issues bar code regulation fact sheet. February 25, 2004 (Available at http://www.fda.gov/oc/initiatives/barcode-sadr/fs-barcode.html [accessed April 13, 2009]).
9. Institute of Safe Medication Practices. ISMP survey shows drug companies providing fewer unit-dosed packaged medications. Medication safety alert. March 6, 2002.

10. VanOnzenoo, H., VandePlas, A., Kessels, A., et al. Factors influencing bar-code verification by nurses during medication administration at a Dutch hospital. *American Journal of Health-System Pharmacy* 2008. 65:644–648.

11. Galvin, L., McBeth, S., Hasdorff, C., et al. Medication bar coding: To scan or not to scan? *Computers, Informatics, Nursing* 2007. 25(2):86–92.

12. Poon, E., Keohane, C., Featherstone, E., et al. Impact of bar code medication administration technology on how nurses spend their time on clinical care. *Journal of Nursing Administration* 2008. 38(12):541–549.

13. Sakowski, J., Newman, J., and Dozier, K. Severity of medication administration errors detected by a bar-code medication administration system. *American Journal of Health-System Pharmacy* 2008. 65:1661–1666.

14. ASHP Foundation. Implementing a bar coded-medication safety program—Pharmacist's tool kit. (Available online: http://www.ashpfoundation.org/MainMenuCategories/Education/SpecialPrograms/ImplemetingaBarCodedMedSafetyProgram.aspx [accessed April 13, 2009]).

15. Institute of Safe Medication Practices. Pathways for medication safety: Assessing bedside bar code readiness. (Available online: http://www.ismp.org/selfassessments/PathwaySection3.pdf [accessed August 23, 2008]).

16. ISMP Medication Safety Alert! If safety is your yardstick, measuring culture from the top down must be a priority. March 22, 2007.

17. Hospital Survey on Patient Safety Culture: 2008 Comparative database report. AHRQ publication no. 08-0039, March 2008. Agency for Healthcare Research and Quality, Rockville, MD. (http://www.ahrq.gov/qual/hospsurvey08/ [accessed August 25, 2008]).

CHAPTER 9

Pharmacy Informatics as a Career

Armen I. Simonian

CONTENTS

9.1 INTRODUCTION

Pharmacy informatics is a relatively new specialty within the practice of pharmacy, creating an exciting career path for the pharmacist who has a passion for computers and ideas for leveraging information technology to support pharmacy activities. This specialty might be well suited to pharmacists who like to tinker with computers, manage a Web site, build a personal computer from components, or program for fun or who are intrigued by the possibility of incorporating these hobbies into their professional practice. One of the most fulfilling job benefits is the satisfaction of knowing that advancing the best and most efficient use of information, aided by deployment of the latest computer technologies, will increase the effectiveness of pharmacy personnel and ultimately improve patient care for a large number of patients. This chapter discusses ways to build a curriculum vitae and job description and to define the typical activities a pharmacist performs in this new specialty practice.

9.2 EDUCATION AND TRAINING

First, the pharmacy informatics specialist should be a pharmacist with a good understanding of pharmacy practice in a variety of practice settings; this experience will help in evaluating the possible applications of information technology in the various stages of the medication use process. An understanding of work flow, coupled with creativity and knowledge of informatics, helps in identifying specific opportunities for increasing efficiency through automation. An understanding of interdisciplinary relationships is also essential for ensuring proper integration of medication management information systems across all areas of healthcare practice.

An effective pharmacy informatics specialist must have an understanding of the goals and regulations of healthcare governing bodies and regulatory agencies. Pharmacy boards and healthcare regulators are now focusing on the proper implementation and monitoring of new technologies. These technologies can alter work flow and practice, and they sometimes introduce failure points, work-arounds, and potential safety issues. Beyond the professional duty to abide by standards of practice and act in the best interest of the patient, the pharmacist specializing in informatics is responsible for ensuring that the informatics tools deployed comply with laws and regulations.

In the course of gaining pharmacy practice experience, the practitioner undoubtedly will learn to use various computer programs, as well as automated packaging and dispensing systems. Expertise in a particular vendor's application is not essential, but the practitioner will benefit from an understanding of the different types of computer programs used in daily practice in both inpatient and outpatient settings. The practitioner should become familiar with general business applications such as word processing, spreadsheet, presentation, and project management software. All of these tools will prove useful in an informatics role.

Work experience can be gained via immediate employment after licensure or by completing a residency program. First-year postgraduate programs provide training in the general aspects of pharmacy practice in health system, managed care, and community settings. Second-year postgraduate programs focus on specific aspects of practice, and the choices today include second-year residencies in pharmacy informatics.

Although a pharmacy degree is a requirement for the pharmacist in an informatics position, a degree in computer science is optional. An understanding of certain computer concepts, though, is essential to be able to evaluate various technologies fully and to interact more effectively with the information systems analysts supporting the pharmacy department. Basic understanding of computer principles, including terminology, data representation, hardware configuration, and software development, is essential. A good educational foundation might include courses covering a general introduction to computers, programming principles, a programming language, operating systems, telecommunications, and database structure.

Because a pharmacy informatics specialist is usually considered a lead or manager position, management experience is highly desirable. This position represents a leadership role with information technology projects, developing and adhering to time lines and coordinating the activities of work teams. At some point in the preparation for this position—specifically when determining the details of the job description and expectations—the scope of practice and responsibilities will have to be defined. Most information management is now computerized.

Thus, theoretically, the responsibilities of the pharmacy informatics specialist could overlap with all aspects of medication use and related processes, including admissions processing, patient billing, departmental financials, document management, scheduling, and other automated aspects of pharmacy operations. Therefore, educational goals could go beyond learning about pharmacy and computers to include topics in healthcare management and business administration.

Once appropriate levels of knowledge and experience are attained, the pharmacy informatics specialist's expertise should be maintained through continuing education. Networking with colleagues is important to be knowledgeable of the systems that are in use and levels of technology implementation, as well as to get ideas on systems that might fit the goals of an institution. One method for staying current and facilitating networking is to attend the major conferences offered by the pharmacy professional societies and healthcare informatics organizations. Examples of these opportunities are the American Society of Health-System Pharmacists (ASHP) Midyear Clinical Meeting, the American Pharmacists Association (APhA) Annual Meeting and Exposition, and the Healthcare Information and Management Systems Society (HIMSS) Annual Conference and Exhibition. Exhibit programs at these conferences provide an opportunity to establish client–vendor partnerships.

Another method for staying current is to subscribe to informatics publications focusing on pharmacy practice, the computer industry in general, and the specific area of healthcare informatics. The *Journal of the American Pharmacists Association* and the *American Journal of Health-System Pharmacy* are examples of pharmacy practice publications. *InformationWeek* covers general business technology topics and is an example of a computer industry publication. As its name implies, *Healthcare Informatics* addresses healthcare information technology and is geared toward an organization's chief information officer. The healthcare informatics societies also offer publications, including the *Journal of Healthcare Information Management, Journal of the American Medical Informatics Association*, and the *Journal of the American Health Information Management Association*.

9.3 JOB OPPORTUNITIES

Job opportunities are most likely to be found in larger hospitals or at the corporate level of multihospital health systems, health maintenance organizations, or other pharmacy chains. The typical management structure of a hospital pharmacy has a pharmacy director. The director may have an assistant or associate director and, at the next level, a number of supervisors such, as an operations manager and a clinical supervisor. In larger hospitals (>400 beds), the pharmacy informatics specialist might hold a full-time position at the supervisor level. In a smaller hospital, the assistant director or operations manager may have to perform multiple functions, including the management of informatics projects.

Multihospital systems, even those composed of small hospitals, can usually justify a pharmacy informatics specialist. The costs of managing informatics projects separately at each hospital can be combined to create a full-time position at the corporate level. The corporate pharmacy informatics specialist would take responsibility for implementation and support of technologies for all the hospital pharmacies within the corporation.

In the retail setting, the independent store owner or pharmacist-in-charge typically takes responsibility for managing the computer system and automation. These individuals rely more on the computer system vendors to provide implementation, maintenance, and enhancement support. Individual stores are generally unable to justify a full-time pharmacy informatics position. Chain pharmacies, though, usually have the needs and resources to hire a full-time pharmacy informatics specialist at the corporate level.

Most hospitals, hospital systems, and chain pharmacies have an information systems (or information technology) department. The information systems department management structure is typically composed of a chief information officer, followed by directors of various divisions such as hardware, programming, networks, and data center. Each director usually has managers who oversee the analysts who perform the work of repairing computers and printers, staffing the help desk, writing and testing programs and databases, and maintaining the Web site and network.

Although much of the pharmacy informatics specialist's work time is spent interacting with information systems department administrators and analysts, the pharmacist usually reports through the pharmacy service's organizational structure. The informatics specialist can best accomplish the role of advocate for pharmacy services informatics initiatives by reporting to the director of pharmacy and sharing the performance goals of the director. The pharmacy informatics specialist should be viewed as a clinician, rather than an analyst, and should also be viewed as a key customer of the information systems department.

9.4 AREAS OF INVOLVEMENT

Principles of specialty practice in pharmacy informatics are delineated in a statement published by ASHP.[1] These principles can be woven into a list of measurable standards, duties, and responsibilities to create the pharmacy informatics specialist job description. In general, the specialist will be directly or indirectly responsible for evaluation, selection, training, configuration, testing, implementation, maintenance, enhancement, and support of all pharmacy information technology. Depending on the scope of the position, institutional structures, and organizational needs and goals, the pharmacy informatics specialist job description might contain some or all of the components listed in Appendix 9.1.

After attaining a position as pharmacy informatics specialist, think about the medication use process and work flow in the practice setting. The specialist should identify all current systems and potential opportunities for the optimal use of information and technologies to promote efficiency and better patient care. He or she should keep in mind the needs of colleagues in medicine, nursing, and other disciplines to make sure that the specialist's ideas and recommendations are in line with the organization's goals. Experience and creativity should be used to develop new programs beyond the informatics applications listed here:

- supply chain
  - electronic ordering and invoicing of wholesaler drug orders
  - just-in-time replenishment system for automated dispensing cabinets

- medication ordering
  - electronic health record (EHR)
  - computerized provider order entry (CPOE)
  - facsimile and digital image transmission of written orders
  - electronic drug information sources
  - decision support deployment, including rules and alerts
- drug dispensing
  - robotic initial dose preparation and cart fill system
  - robotic intravenous solution preparation
  - automated dispensing cabinets (ADCs)
  - parenteral nutrition admixture compounding pumps
- drug administration
  - electronic medication administration record (e-MAR)
  - bar code medication administration (BCMA)
  - computer-based infusion devices (smart pumps)
- drug monitoring
  - pharmacokinetic dosing tools
  - electronic documentation of clinical interventions
  - electronic documentation of adverse drug events and
  - automated reports that trigger follow-up

9.4.1 Project Management

Project management is a large part of the informatics specialist's responsibilities. Selection and implementation of projects are done in collaboration with a number of individuals. Many useful tools are available to document and track these projects. Simple lists and grids can be built using word processing or spreadsheet software. Project management software suites, such as Microsoft Project (Microsoft Corporation, Redmond, WA), provide more sophisticated features for managing projects and both human and financial resources. Project tasks can be documented, time lines formulated, milestones set, and responsibilities and resources allocated. It is important to use the documentation procedures specified by the organization, particularly with respect to the organization's information technology governance and change-management processes.

The specialist should look for opportunities to standardize applications across the organization and develop implementation standards for each application. For all applications that are implemented, he or she should develop and adhere to policies and procedures for the safe and effective use of the software and technologies. Beyond proper use, numerous other aspects of computer applications are necessary:

- maintaining and monitoring the applications

- addressing standard software settings

- assessing employee education and competence

- addressing employee access and login security

- backing up and archiving data and

- developing planned and unanticipated downtime procedures

For consistency across applications and organizational entities, drug database mnemonic and nomenclature standards should be created and adhered to. Designations for every product, such as an intravenous drug solution, should be consistent across all systems from ordering to administration to the patient. Collaboration with nursing and other disciplines is necessary to define standard products available for ordering and agree on standard dose-administration times. These types of standardization efforts add to the safety of clinical systems.

9.4.2 Personnel Management

Once the informatics specialist has established his or her position, a pharmacy services informatics team should be built. Each entity (hospital or retail pharmacy) of a multisite institution should have a representative on the team. In the pharmacy, this staff member is usually a pharmacist who has responsibility for submitting change management requests and providing local informatics support. These individuals might be called "superusers" or "database managers," and they may participate in projects as time allows. Organizations might also have pharmacy technicians who specialize in automated dispensing, and they also should be part of the informatics team. Team meetings help with information sharing, resolution of issues, and project planning.

On a more global basis, regular meetings should be held to address organizational pharmacy informatics topics. The goals of the pharmacy informatics group meetings are to make decisions on system functionality, implementation, work flow, quality, safety, and standardization. The appropriate membership of internal and external representatives who can address interrelationships between pharmacy applications and other systems should be included. Regular attendees should include the pharmacy information specialist, pharmacy database managers, pharmacy directors, information systems pharmacy lead and analysts, and nursing and physician informatics specialists. Content experts, such as clinical pharmacy and medication safety specialists, and experts from other disciplines may be invited as needed to address specific issues. The group sets priorities for changes and enhancements. Items that have a direct impact on other systems and disciplines are referred to the appropriate groups representing those disciplines.

To build vendor relationships and provide more effective support for specific technologies, separate meetings should be held for teams that address a particular technology. For example, one team might focus on automated dispensing cabinets, and the core membership for this group might include the pharmacy informatics specialist, technician superusers, information systems pharmacy analysts, and vendor representatives. Ad hoc membership could include pharmacy directors and information systems leads.

9.5 SUMMARY

Specific roles and responsibilities of the pharmacy informatics specialist are still being defined, and the job description will probably change over time. A creative individual with knowledge of pharmacy and computer science has tremendous opportunities to improve the efficiency, safety, and effectiveness of medication use processes through the appropriate application of information technology.

REFERENCES

1. ASHP statement on the pharmacist's role in informatics. *American Journal of Health-System Pharmacists* 2007. 64:200–203.

APPENDIX 9.1: INFORMATICS SPECIALIST JOB DESCRIPTION COMPONENTS

1. Advocate for pharmacy services information technology goals and objectives. Promote the best use of information aided by deployment of the latest computer technologies to support and improve operational and clinical effectiveness.

2. Act as a liaison between the pharmacy service, information systems, and application vendors. Establish and maintain positive working relationships, using appropriate language and terminology to communicate technical issues to nontechnical customers.

3. Work with pharmacy staff and administration and with the information systems department to determine and assess the information technology needs of pharmacy and its customers, both patients and staff within the institution.

4. Continually assess new technology that may assist the pharmacy staff in providing quality clinical and distributive care. Serve as a resource for questions about new and existing pharmacy information technologies.

5. Develop objective measures for evaluating competing products within a particular category of pharmacy informatics applications.

6. Display foresight and creativity in suggesting new technologies that improve patient care.

7. Participate in the organization's change management process, collaborating with the information systems department to document and implement change requests appropriately.

8. Take into account the principles of interdisciplinary integration and alignment with corporate goals when considering new technologies and evaluating the validity of currently installed systems.

9. Participate in multientity or multidisciplinary teams, focus groups, task forces, and other designated activities that further the development of optimal pharmacy information systems and the practice of pharmacy.

10. Create and ensure adherence to drug database build standards, including mnemonic and naming conventions.

11. Maintain the integrity of medication databases and related tables (e.g., frequencies, routes, dose forms, units of measure). Ensure synchronization of databases and tables across pharmacy, EHR, e-MAR, BCMA, ADC, and infusion pump software.

12. Participate in the creation of test plans and coordinate the validation of all database builds, program settings, and system interfaces.

13. Coordinate a matrix of several simultaneous projects using appropriate prioritization, timing, and creativity to ensure that projects are completed on time and within resource constraints.

14. Develop policies, procedures, and departmental guidelines for the safe and effective use of technologies, including configuration, quality assurance, training, access, downtime, backup, and archiving.

15. Promote the best use of safety functionality (e.g., clinical decision support information, rules, and warnings) to improve medication use safety.

16. Document and investigate issues and errors reported by pharmacy application users and work with information systems and the pharmacy application vendors to repair, enhance, and upgrade currently installed systems.

17. Create and provide data download analysis, documents, and reports, as needed, by pharmacy administration and staff.

18. Establish appropriate charge formulas, ensure proper coding of drug database entries, and audit periodic pricing updates.

19. Establish daily billing activities for editing, crediting, and monitoring medication charge transactions. Participate in internal and external financial audits of patients' medication charges.

20. Investigate reports of potential and actual adverse drug events attributable to information technology, address root causes, eliminate opportunities for similar future errors, and monitor the effectiveness of correction plans.

21. Evaluate the impact of automated pharmacy methodologies on the legal, financial, ethical, and managerial principles set forth by the organization.

22. Ensure compliance of all pharmacy information technologies with goals, standards, laws, and regulations of healthcare governing bodies and regulatory agencies.

23. Review healthcare literature, maintain pharmacy and informatics professional organization affiliations, and network with colleagues at external institutions to maintain awareness of best practices, system experience, and implementation of the latest technologies.

24. Perform ongoing review of needs related to pharmacy informatics education and training. Assist with the development and presentation of needs-based education, including update and refresher courses for pharmacists, nurses, physicians and other healthcare personnel.

25. Teach and mentor pharmacy students and residents to impart knowledge of pharmacy informatics as a specialty area of pharmacy practice, working closely with schools of pharmacy and residency coordinators to ensure adherence to teaching guidelines, learning objectives, and documentation.

CHAPTER 10

Avoiding Medication Errors

Joseph E. Scherger and Grace M. Kuo

CONTENTS

10.1 INTRODUCTION

Errors in medical practice are unfortunately very common. The Institute of Medicine (IOM) report, *To Err Is Human,*[1] stated that between 44,000 and 98,000 people in the United States die every year due to medical errors. This puts medical errors between the fourth and eighth leading causes of death in the country. A high proportion of these medical errors are related to medications. The IOM report used data only from hospitals; combining these with errors in all sites of medical practice, the death rate from errors in medicine would greatly exceed 100,000 deaths each year.

In a follow-up IOM report, *Preventing Medication Errors,* released in July 2006, at least 1.5 million preventable adverse drug events were estimated to occur each year in the United States.[2] According to this report, 380,000–450,000 preventable adverse drug events occur in hospital settings, 800,000 preventable adverse drug events take place in long-term-care facilities, and at least 530,000 preventable adverse drug events are estimated among outpatient Medicare patients in ambulatory care settings. Because these statistics are derived from voluntary reports, the actual number of adverse drug events is most likely underestimated.

This chapter focuses on the role of informatics in reducing medication errors. Computer technologies alone have a role in medication error prevention if applied wisely. But technology alone does not provide safety and may even cause harm. The interplay between technology and healthcare professionals is crucial in the development of patient safety systems. Three key professionals are involved with medication administration: physicians, pharmacists, and nurses. Numerous studies show that when these professionals work interdependently as a team, the checks and balances greatly enhance patient safety.[3–7]

Adding the patient and family to the medication administration process also enhances safety.[8,9] Hospitals, community-based clinics, office practices, long-term-care facilities, and the patient's home are all discussed here with a focus on optimizing safe medication practices. The primary focus in this chapter is on the role of the pharmacist as a critical member of the team interacting with physicians, nurses, and patients and using modern information technology to achieve optimal medication practices.

10.1.1 The Healthcare Team

In the traditional culture in medical practice, the physician has ultimate authority and practices with professional autonomy. This culture also has it that pharmacists and nurses are "ancillary" professionals that carry out the orders of the physician. This traditional culture is neither safe nor effective. Without disparaging physicians and their central role in patient care, pharmacists and nurses must be empowered to participate with authority on the professional team. For example, one situation that exemplifies true teamwork is whether the pharmacist or the nurse can stop a physician order for a safety concern. Optimal systems of care must allow for and even encourage such professional authority among pharmacists and nurses. Even if the concern is a false alarm, pharmacists and nurses should be rewarded for stepping forward in optimal healthcare systems.

Optimal systems of healthcare for medication safety have two characteristics: teamwork among the professionals, staff, and patients and excellent information systems. This combination of human factors and informatics maximizes the potential to deliver the safest and highest quality care. Having all patients and professionals working together from a common information system with built-in safety technology has the potential to achieve unprecedented safety and should be the hallmark of any modern medication administration system, regardless of setting. Optimal systems of care have the pharmacists, nurses, and physicians meeting regularly to go over errors and "near misses" and refining medication administration processes.

10.2 MEDICATION ERRORS IN VARIOUS SETTINGS

10.2.1 Medication Errors in Hospitals

Medication errors in hospitals are common and have the potential for causing serious harm. Most hospitalized patients are on intravenous (IV) medications where errors result in risk of serious injury and death. As stated earlier, *To Err Is Human* used hospital data and reported on patient safety problems that resulted in 44,000–98,000 preventable deaths a year. Medication errors played a major role in these deaths and in harm to patients that did not result in death. From the studies covered by this IOM report, 7% of all patients admitted to the hospital experience a clinically important medication error.[1] In studies done in the 1990s, the average cost of these medication errors was $4,700 per admission.[1]

A limited number of medications dominate the list of drugs involved with major medication errors in hospitals. These include insulin, heparin, warfarin, and narcotic analgesics. Often, a lack of standardization contributes to harm because protocols differ among physicians. If physicians act alone in ordering medications, followed without question by a nurse's transmission of the orders and a pharmacist's filling the orders, physician errors are not recognized or questioned. Standardization of drug prescribing and administration is a critical first step in reducing medication errors.[10] Information technology and teamwork among the professionals can then interplay to create safe systems of medication administration.[11]

Hospital medication errors continue to capture headlines. For example, in two of the nation's leading hospitals, newborn babies were given an adult dose of heparin due to a mix-up in heparin vials. In one of the hospitals, four babies died of bleeding complications. Tragically, one hospital did not learn from the other: These events happened a year apart. The following is the assessment of the Institute for Safe Medication Practices[12]:

> Recently, an error occurred at a hospital in which 10,000 units/mL heparin vials were used to flush the vascular access lines of infants rather than the 10 units/mL vials. Previously, a similar error had occurred at a different hospital when the 10,000 units/mL vials were stocked mistakenly in the space reserved for the 10 units/mL vials. After the occurrence of the first error, several recommendations were made in order to prevent a recurrence of this type of error such as separating the 10,000 unit/mL vials from all other strengths, verifying all drugs taken to be stocked in patient care areas, implementing bar-code, and using pre-filled heparin flush syringes instead of vials. Despite these recommendations, the second heparin error still occurred. This demonstrates two issues plaguing the healthcare industry that cause concern for patient safety.
>
> - The first issue is that most healthcare organizations are not doing enough to educate themselves about potential risks and errors existing both within the organization and externally. Many resources are available, including the ISMP newsletters, but because organizations rarely seek out this information and make recommended changes accordingly, the mistakes of others are often repeated.

- The second issue is that healthcare organizations are not proactive enough when it comes to evaluating their own systems and procedures for the possibility of potential errors.

Errors in pediatric hospitals are at least as common as in adult hospitals and are often more serious due to the fragile nature of sick children, especially neonates.[13-16] Medication administration in children often requires calculations based on weight and other variables; standardization and information technology that performs critical calculations allow for much greater safety.[17]

10.2.2 Medication Errors in Community Clinics and Medical Offices

Studies in community clinics and medical office practice indicate that outpatient medication prescribing errors occur in between 7.8 and 21% of patients and adverse drug events occur in 18 and 25% of patients.[2,18,19] Examples of the reported prescribing errors include inappropriate medication selection, omitting necessary information on the prescription, selecting incorrect dose or directions, unclear quantity to be dispensed, and potential adverse drug–drug interactions. Examples of adverse drug events include adverse reactions caused by a medication error or a previously documented causative drug. Medication discrepancies are prevalent among ambulatory care clinics, ranging from 26 to 76%.[19] Examples of the medication discrepancies include medications that patients are taking that are not recorded in the chart or medications that are recorded in the chart that patients are not taking. Furthermore, medication errors from inadequate therapeutic monitoring commonly occur.[19]

Over three billion prescriptions were sold in U.S. outpatient and community pharmacies in 2006; retail prescription-drug expenditures totaled $275 billion.[20] As medication use increases, the risk of medication errors also increases. One cross-sectional study conducted in 50 community pharmacies located in six U.S. cities found the dispensing error rate to be approximately four errors per 250 prescriptions per pharmacy per day.[21] Extrapolating from this finding, more than 50 million medication errors could occur from filling three billion prescriptions each year. Studies show that serious harm may come from these errors and that they are preventable.[18,22]

10.2.3 Medication Errors in Long-Term Care

Medication errors are common in long-term-care facilities, especially in nursing homes.[23,24] Historically, the transmission of medication information from hospitals to nursing homes has been poor. Medication allergies and changes in medications are often missing in paper-based systems that require human transfers of information. The patient and family are often not available to consult about medications the patient takes, including the exact name, dose, or frequency of medications used by the patient. Medication reconciliation among facilities and providers is critical and is one of the most important strategies for new health information systems.[7]

Lapses in medication monitoring are thought to be the most common type of error in the nursing home setting, although only a limited number of studies evaluating errors in this setting have been conducted. The frequency of errors associated with medication administration is estimated to range between 6 and 20 errors per 100 doses.[2]

10.2.4 Medication Errors in the Home

At home, medication errors occur when patients take the wrong drug or wrong dose or take the right drug and dose at the wrong time.[25] In a study of 6,718 elderly home-care patients, 30% had potential medication errors when either the Beers criteria or the home health criteria were applied.[26] The Beers criteria are widely used to "identify patterns of medication use that unnecessarily place older persons at risk of adverse drug reactions."[26] The home health criteria were developed to "identify home healthcare patients whose patterns of medication use and signs and symptoms provided sufficient evidence of risk of a clinically important adverse drug effect to warrant reassessment of the patient."[26]

Errors also occur when patients do not take their medications. Medication noncompliance is estimated to cost $100 billion in the United States each year and is becoming a public health concern.[27,28] The community pharmacist is in an ideal position to monitor medication use in the home and provide education to patients and families. A robust information system that helps pharmacists review patients' medications and provides an interactive communication tool with patients' physicians (e.g., http://MedActionPlan.com) can greatly facilitate this process.[3,29]

10.3 CAUSES OF MEDICATION ERRORS

Medication errors are caused by many factors. *To Err Is Human* was so named because it is normal for human beings to make mistakes. The culture of healthcare has traditionally held the belief that physicians, nurses, and pharmacists are so well educated that they would not make mistakes if they were careful. This is an inappropriate and dangerous belief. Human beings, no matter how well educated, make mistakes 2–5% of the time, especially when doing repetitive tasks. The error rate increases when the professionals are tired or hurried.[30]

10.3.1 Root Cause Analysis

Root cause analysis (RCA) is a formal method used to investigate an error.[30–32] RCA is used by industry but has only recently been applied in healthcare settings. Performing an RCA involves an individual or a team analyzing all the steps leading to an error event. Three types of factors are analyzed: human, organizational, and technical. Human factors involving one or more persons causing the error can be uncovered through nonthreatening interviews. Everyone involved in the error is interviewed. Organizational factors include structural problems, such as an insufficient staff, inadequate supervision, or inadequate training for a task. Technical factors include having to work with malfunctioning equipment. Once all of these factors are analyzed in detail, a report is written about the RCA with recommendations for improvement. The final critical step with any RCA is to ensure that the recommendations for improvement are put in place and maintained.

10.4 TECHNOLOGIES THAT ENHANCE MEDICATION SAFETY

Three emerging information technologies are being studied for their potential to reduce medication errors: computerized provider order entry, bar coding, and electronic prescribing. These specific technologies may exist separately or as part of a common informatics platform, the electronic health record (EHR). Pharmacists play a key role in the selection, adoption, and successful application of these technologies in any setting (see Chapters 7 and 9). These technologies, decision support programs, and automated dispensing systems can reduce rates of medication errors.[11,33] However, technologies can also cause errors, so diligence in their application is critical, and pharmacists play a major role in monitoring their use. Each technology is discussed separately next.

10.4.1 Computerized Order Entry

Often called computerized provider order entry (CPOE), this technology requires the person placing an order for a patient to do so using an information system platform rather than hand-writing or verbally requesting an order. CPOE is one of the most studied new information technologies and its success in reducing medication errors has been mixed. On one hand, some CPOE systems have resulted in dramatic reductions in medication errors in both pediatric and adult settings.[34–42] On the other hand, CPOE systems have resulted in medication errors (e.g., by allowing providers to select the wrong medication or medication directions from pull-down lists), showing that to err is not always human.[39,43–46] Nevertheless, the influential Leapfrog Group considers CPOE an important quality indicator. Their 2008 survey of hospitals found that only 8% of the institutions surveyed had implemented CPOE.[47]

Information technologies such as CPOE should not be put in place without healthcare professional oversight because the application must make sense in the clinical context and clinicians are vital in identifying misuses of the technology. Pharmacists are in an ideal position to help implement the recommendations and monitor the outcomes of CPOE and are able to identify and avoid potential points of errors. With their expertise in drug information, pharmacists are vital for achieving maximum medication safety.[48] Think of this situation as similar to aviation safety, where the pilot and automatic pilot work together.

10.4.2 Electronic Prescribing

To Err Is Human called for the elimination of handwritten prescriptions and other orders. This has not happened yet, mainly due to the high cost of infrastructure and equipment for implementing systemwide electronic prescribing; however, its implementation is growing rapidly. Electronic prescribing (also called eRx) requires the prescriber to use a computer system that is compliant with standardized eRx features that interface with pharmacy computerized systems. Through eRx, the patient's name, the medication name, and dosage are far more likely to be correct.

Using computer systems in the prescribing of medications offers great potential for implementing safe medication practices. When these systems are well designed and targeted to certain patient populations and clinical conditions, the safety results are impressive.[49–54] Here, again, professional oversight, especially by a pharmacist, ensures that these systems perform properly and are maintained and improved based on feedback and measured clinical experience.

10.4.3 Clinical Decision Support with Safety Features

As eRx, CPOE, and EHRs become more common, efforts are underway to improve the clinical decision support systems (CDSSs) built into these tools. The first CDSS application that has been present for a long time is drug alerting: pop-up messages that are automated into the prescribing process. Computerized decision support tools embedded within CPOE are effective in reducing hospital-based and outpatient medication errors.[55,56] Informatics tools can detect and prevent medication errors and adverse drug events.[57]

High-quality clinical decision support software is crucial in reducing errors (see Chapter 15).[58] Early CPOE systems failed to protect against many medication errors and sometimes caused harm (e.g., due to lack of safeguard features or allowing for free text typing that resulted in spelling errors). Newer systems have a better track record. Also, narrow applications of CPOE, such as in pediatric or neonatal critical care, seem more effective than in more complex situations like general adult medicine, where the patient mix and therapies are more variable.

Prescribing alerts can be highly effective when targeted to specific medications and clinical conditions. For example, calculation of dosages of critical medications such as digoxin in elderly patients has enhanced safety with electronic support.[59,60] Computer signals related to medication safety alerts in Medicare enrollees at multispecialty group practices identified 53% of potential incidents.[61] In a prospective analysis after one hospital implemented CPOE, the medication error rate (excluding missed doses) fell by 81% and nonintercepted serious medication errors fell by 86%.[34] Though CPOE is generally beneficial, system "noise" from nonstandardized alerts and signals may be annoying and not helpful in reducing errors; therefore, standardized alerts and electronic prescribing standards are still needed to improve medication safety. However, if the alert threshold is set too low or implementation is poor, many drug alerts will be inappropriate or unhelpful.[19,62,63]

Physicians' and pharmacists' almost universal experience with pop-up drug alerts has been that very few of the alerts are clinically relevant to the patient being treated. The alerts are often triggered at such a low threshold that almost meaningless information comes up. This results in "alert fatigue" in which the physician or pharmacist ignores or bypasses the alert without even reading it or giving it serious thought. Alert fatigue is dangerous because the occasional alert that is important is likely to be missed. Effective clinical decision support tools must present alerts that are clinically important most of the time and not burden the professional with meaningless messages.[62,64] Clinical decision support tools are in their early stages and have the potential for many more applications that will improve both the safety and overall quality of medical practice. (See Chapter 15 for a more complete discussion.)

10.4.4 Bar Coding

Bar coding is used in many industries to prevent human errors of recording and calculation and to ensure proper identification. Increasingly, patients are being assigned a bar coded wrist band when they are admitted to a hospital or other healthcare facility. All medication and other orders such as requisitions for laboratory tests and x-rays are also bar coded to ensure that the right patient receives the correct intervention. Bar coding reduces 60–80% of administration errors[65–67]; however, as with any technology, careful oversight is important to avoid the potential for bar coding to cause errors.[68] (See Chapter 8 for a more complete discussion.)

10.4.5 Other Tools

Other electronic tools, such as medication error reporting programs using computerized forms, allow healthcare professionals to document errors that have occurred in their practice setting. The reported errors, including prescribing, dispensing, administering, and monitoring errors and the type of medications involved, can inform the healthcare organization where errors occur and what areas need improvement.[69] In addition, automated diagnostic and pharmacy data systems can be used for surveillance purposes to track illnesses and assess medication use (see Chapter 16).[70]

An innovative method that links an ambulatory care EHR and a Web-based patient portal has been implemented for patients with diabetes mellitus. This has allowed physicians to make medication dosage adjustments in a timely manner and encouraged patients to be more engaged with their care plans.[71] Other disease management tools and patient registries have also been successfully implemented in the EHR to enhance chronic disease management and improve the quality of care.[72,73]

10.5 PHARMACISTS' ROLES IN MEDICATION SAFETY

The contributions of pharmacists in preventing medication-related problems for adult and pediatric patients in both inpatient and outpatient settings is well documented.[3,6,29,74–79] For example, for patients seen in adult outpatient clinics, pharmacist identification of drug-related problems through medical record review has helped prevent adverse consequences.[8] For patients being discharged from the hospital, pharmacist counseling has helped prevent adverse drug events after they leave the hospital.[80] However, the current pharmacist shortage is having a negative impact on medication errors.[81]

The role of the pharmacist is expanding with the advancement of informatics technology and the increasing demand for electronic health record systems to continue efforts of medication error prevention. Increasingly, pharmacists are needed to contribute their expertise in drug management through designing and implementing healthcare informatics tools. In addition to their skills in dispensing and drug monitoring, pharmacists can help develop informatics decision-making tools with medication safety features designed to prevent errors.[82] Computerized tools, in turn, enable pharmacists to expand their consultation potential and improve the quality of healthcare provided to patients.[83] In one study, 78% of potentially harmful prescribing errors were intercepted by pediatric pharmacists using CPOE.[84]

10.6 SUMMARY

This chapter has explored many of the informatics applications that can enhance medication safety. Optimal healthcare systems are committed to having the best information systems available. For example, we know that the quality of CPOE systems depends on how refined the clinical decision support system is within it. Systems for bringing consistent knowledge to the point of care are continuously improving and need to be upgraded. Finally, no information systems should function alone, and the pharmacist well trained in informatics is ideally suited to provide leadership and oversight in the application of these lifesaving technologies.

REFERENCES

1. Kohn, L. T., Corrigan, J. M., and Donaldson, M. S. *To err is human: Building a safer health system.* Institute of Medicine. Washington, D.C.: National Academy Press, 1999.
2. Aspden, P., Wolcott, J., Bootman, J. L., and Cronenwett, L. R. *IOM report: Preventing medication errors.* Institute of Medicine. Washington D.C.: National Academy Press, 2006.
3. Brown, C. A., Bailey, J. H., Lee, J., Garrett, P. K., and Rudman, W. J. The pharmacist–physician relationship in the detection of ambulatory medication errors. *American Journal of Medical Science* 2006. 331:22–24.
4. Franklin, B. D., O'Grady, K., Paschalides, C., et al. Providing feedback to hospital doctors about prescribing errors; a pilot study. *Pharmacy World & Science* 2007. 29:213–220.
5. Kallail, K. J., and Stanton, S. R. Pharmacy-physician communications: Potential to reduce medication errors. *Journal of the American Pharmacists Association* 2006. 46:618–620.
6. Leape, L. L., Cullen, D. J., Clapp, M. D., et al. Pharmacist participation on physician rounds and adverse drug events in the intensive care unit. *JAMA* 1999. 282:267–270.
7. Varkey, P., Cunningham, J., O'Meara, J., Bonacci, R., Desai, N., and Sheeler, R. Multidisciplinary approach to inpatient medication reconciliation in an academic setting. *American Journal of Health-System Pharmacy* 2007. 64:850–854.
8. Viktil, K. K., Blix, H. S., Moger, T. A., and Reikvam, A. Interview of patients by pharmacists contributes significantly to the identification of drug-related problems (DRPs). *Pharmacoepidemiology and Drug Safety* 2006. 15:667–674.
9. Weingart, S. N., Toth, M., Eneman, J., et al. Lessons from a patient partnership intervention to prevent adverse drug events. *International Journal of Quality Health Care* 2004. 16:499–507.
10. Donihi, A. C., DiNardo, M. M., DeVita, M. A., and Korytkowski, M. T. Use of a standardized protocol to decrease medication errors and adverse events related to sliding scale insulin. *Quality Safety Health Care* 2006. 15:89–91.
11. Bates, D. W. Using information technology to reduce rates of medication errors in hospitals. *British Medical Journal* 2000. 320:788–791.
12. Institute for Safe Medication Practices. Another heparin error: Learning from mistakes so we don't repeat them (http://www.ismp.org/Newsletters/acutecare/articles/20071129.asp).
13. Chedoe, I., Molendijk, H. A., Dittrich, S. T., et al. Incidence and nature of medication errors in neonatal intensive care with strategies to improve safety: A review of the current literature. *Drug Safety* 2007. 30:503–513.
14. Kaushal, R., Bates, D. W., Landrigan, C., et al. Medication errors and adverse drug events in pediatric inpatients. *JAMA* 2001. 285:2114–2120.
15. Kozer, E., Berkovitch, M., and Koren, G. Medication errors in children. *Pediatric Clinics of North America* 2006. 53:1155–1168.
16. Simpson, J. H., Lynch, R., Grant, J., and Alroomi, L. Reducing medication errors in the neonatal intensive care unit. *Archives of Disease in Childhood—Fetal and Neonatal Edition* 2004. 89:F480–F482.
17. Jacobs, B. Electronic medical record, error detection, and error reduction: A pediatric critical care perspective. *Pediatric Critical Care Medicine* 2007. 8(2 Suppl):S17–S20.

18. Gandhi, T. K., Weingart, S. N., Seger, A. C., et al. Outpatient prescribing errors and the impact of computerized prescribing. *Journal of General Internal Medicine* 2005. 20:837–841.

19. Kuo, G. M. Medication errors in community/ambulatory care: Incidence and reduction strategies. *Journal of Pharmaceutical Finance and Economic Policy* 2006. 15:43–136.

20. MS Reports U.S. prescription sales jump 8.3 percent in 2006, to $274.9 billion: Managed Care Business Week (http://www.newsrx.com/newsletters/Managed-Care-Business-Week/2007-03-27/80327200794MB.html 2007).

21. Flynn, E. A., Barker, K. N., Gibson, J. T., Pearson, R. E., Berger, B. A., and Smith, L. A. Impact of interruptions and distractions on dispensing errors in an ambulatory care pharmacy. *American Journal of Health-System Pharmacy* 1999. 56:1319–1325.

22. Gurwitz, J. H., Field, T. S., Harrold, L. R., et al. Incidence and preventability of adverse drug events among older persons in the ambulatory setting. *JAMA* 2003. 289:1107–1116.

23. Gurwitz, J. H., Field, T. S., Avorn, J., et al. Incidence and preventability of adverse drug events in nursing homes. *American Journal of Medicine* 2000. 109:87–94.

24. Pierson, S., Hansen, R., Greene, S., et al. Preventing medication errors in long-term care: Results and evaluation of a large scale web-based error reporting system. *Quality Safety Health Care* 2007. 16:297–302.

25. Mager, D. R. Medication errors and the home care patient. *Home Healthcare Nurse* 2007. 25:151–155.

26. Meredith, S., Feldman, P. H., Frey, D., et al. Possible medication errors in home healthcare patients. *Journal of the American Geriatric Society* 2001. 49:719–724.

27. Cleemput, I., and Kesteloot, K. Economic implications of non-compliance in healthcare. *Lancet* 2002. 359(9324):2129–2130.

28. Cleemput, I., Kesteloot, K., and DeGeest, S. A review of the literature on the economics of non-compliance. Room for methodological improvement. *Health Policy* 2002. 59:65–94.

29. Martin, C. M., and Bryan, G. Pharmacists at the forefront: Reducing medication errors. *Consultant Pharmacist* 2006. 21:380–384, 387–389.

30. Reason, J. *Human error.* New York: Cambridge University Press, 1990.

31. Brown, M., Frost, R., Ko, Y., and Woosley, R. Diagramming patients' views of root causes of adverse drug events in ambulatory care: An online tool for planning education and research. *Patient Education and Counseling* 2006. 62:302–315.

32. Wu, A. W., Lipshutz, A. K. M., and Pronovost, P. J. Effectiveness and efficiency of root cause analysis in medicine. *JAMA* 2008. 299:685–687.

33. Bates, D. W., Cohen, M., Leape, L. L., Overhage, J. M., Shabot, M. M., and Sheridan, T. White paper: Reducing the frequency of errors in medicine using information technology. *Journal of the American Medical Informatics Association* 2001. 8:299–308.

34. Bates, D. W., Teich, J. M., Lee, J., et al. The impact of computerized physician order entry on medication error prevention. *Journal of the American Medical Informatics Association* 1999. 6:313–621.

35. Cordero, L., Kuehn, L., Kumar, R. R., and Mekhjian, H. S. Impact of computerized physician order entry on clinical practice in a newborn intensive care unit. *Journal of Perinatology* 2004. 24:88–93.

36. Jayawardena, S., Eisdorfer, J., Indulkar, S., Pal, S. A., Sooriabalan, D., and Cucco, R. Prescription errors and the impact of computerized prescription order entry system in a community-based hospital. *American Journal of Therapeutics* 2007. 14:336–340.

37. Kim, G. R., Chen, A. R., Arceci, R. J., et al. Error reduction in pediatric chemotherapy: Computerized order entry and failure modes and effects analysis. *Archives of Pediatric and Adolescent Medicine* 2006. 160:495–498.

38. Potts, A. L., Barr, F. E., Gregory, D. F., Wright, L., and Patel, N. R. Computerized physician order entry and medication errors in a pediatric critical care unit. *Pediatrics* 2004. 113:59–63.

39. Shulman, R., Singer, M., Goldstone, J., and Bellingan, G. Medication errors: A prospective cohort study of hand-written and computerized physician order entry in the intensive care unit. *Critical Care* 2005. 9:R516–R521.

40. Teich, J. M., Merchia, P. R., Schmiz, J. L., Kuperman, G. J., Spurr, C. D., and Bates, D. W. Effects of computerized physician order entry on prescribing practices. *Archives of Internal Medicine* 2000. 160:2741–2747.

41. Upperman, J. S., Staley, P., Friend, K., et al. The impact of hospital-wide computerized physician order entry on medical errors in a pediatric hospital. *Journal of Pediatric Surgery* 2005. 40:57–59.

42. Vardi, A., Efrati, O., Levin, I., et al. Prevention of potential errors in resuscitation medication orders by means of a computerized physician order entry in pediatric critical care. *Resuscitation* 2007. 73:400–406.

43. Horsky, J., Kuperman, G. J., and Patel, V. L. Comprehensive analysis of a medication dosing error related to CPOE. *Journal of the American Medical Informatics Association* 2005. 12:377–382.

44. Koppel, R., Metlay, J. P., Cohen, A., et al. Role of computerized physician order entry systems in facilitating medication errors. *JAMA* 2005. 293:1197–1203.

45. Walsh, K. E., Adams, W. G., Bauchner, H., et al. Medication errors related to computerized order entry for children. *Pediatrics* 2006. 118:1872–1879.

46. Zhan, C., Hicks, R. W., Blanchette, C. M., Keyes, M. A., and Cousins, D. D. Potential benefits and problems with computerized prescriber order entry: Analysis of a voluntary medication error-reporting database. *American Journal of Health-System Pharmacy* 2006. 63:353–358.

47. Leapfrog Group Hospital Survey finds majority of hospitals fail to meet important quality standards (http://www.leapfroggroup.org/media/file/2008_Survey_results_final_042909.pdf).

48. Bobb, A., Gleason, K., Husch, M., Feinglass, J., Yarnold, P. R., and Noskin, G. A. The epidemiology of prescribing errors: The potential impact of computerized prescriber order entry. *Archives of Internal Medicine* 2004. 164:785–792.

49. Bergeron, B. Medication errors and e-prescribing: Solutions and limitations. *Journal of Medical Practice Management* 2004. 20:152–153.

50. Huertas-Fernandez, M. J., Baena-Canada, J. M., Baena-Cañada, J. M., Martínez-Bautista, M. J., Arriola-Arellano, E., and García-Palacios, M. V. Impact of computerized chemotherapy prescriptions on the prevention of medication errors. *Clinical and Translational Oncology* 2006. 8:821–825.

51. Lenderink, B. W., and Egberts, T. C. Closing the loop of the medication use process using electronic medication administration registration. *Pharmacy World & Science* 2004. 26:185–190.

52. Mirco, A., Campos, L., Falcão, F., Nunes, J. S., and Aleixo, A. Medication errors in an internal medicine department. Evaluation of a computerized prescription system. *Pharmacy World & Science* 2005. 27:351–352.

53. Schiff, G. D., and Rucker, T. D. Computerized prescribing—Building the electronic infrastructure for better medication usage. *JAMA* 1998. 279:1024–1029.

54. Wu, R. C., Laporte, A., and Ungar, W. J. Cost-effectiveness of an electronic medication ordering and administration system in reducing adverse drug events. *Journal of Evaluation in Clinical Practice* 2007. 13:440–448.

55. CPOE study shows drop in hospital errors. May, 2006. *Healthcare Benchmarks Quality Improvement* 13:54–55.

56. Hospital uses CPOE system to reduce medication errors. June, 2006. *Performance Improvement Advisory* 10:49–52.

57. Anderson, J. G. Information technology for detecting medication errors and adverse drug events. *Expert Opinions in Drug Safety* 2004. 3:449–455.

58. Kaushal, R., Shojania, K. G., and Bates, D. W. Effects of computerized physician order entry and clinical decision support systems on medication safety—A systematic review. *Archives of Internal Medicine* 2003. 163:1409–1416.

59. Galanter, W. L., Polikaitis, A., and DiDomenico, R. J. A trial of automated safety alerts for inpatient digoxin use with computerized physician order entry. *Journal of the American Medical Informatics Association* 2004. 11:270–277.

60. Simon, S. R., Smith, D. H., Feldstein, A. C., et al. Computerized prescribing alerts and group academic detailing to reduce the use of potentially inappropriate medications in older people. *Journal of the American Geriatric Society* 2006. 54:963–968.

61. Field, T. S., Gurwitz, J. H., Harrold, L. R., et al. Strategies for detecting adverse drug events among older persons in the ambulatory setting. *Journal of the American Medical Informatics Association* 2004. 11:492–498.

62. Spina, J. R., Glassman, P. A., Belperio, P., Cader, R., and Asch, S. Clinical relevance of automated drug alerts from the perspective of medical providers. *American Journal of Medical Quality* 2005. 20:7–14.

63. Weingart, S. N., Toth, M., Sands, D. Z., Aronson, M. D., Davis, R. B., and Phillips, R. S. Physicians' decisions to override computerized drug alerts in primary care. *Archives of Internal Medicine* 2003. 163:2625–2631.

64. Feldstein, A., Simon, S. R., Schneider, J., et al. How to design computerized alerts to safe prescribing practices. *Joint Commission Journal on Quality and Safety* 2004. 30:602–613.

65. Anderson, S., and Wittwer, W. Using bar-code point-of-care technology for patient safety. *Journal of Healthcare Quality* 2004. 26:5–11.

66. Goth, G. Raising the bar. Bar coding has the potential to dramatically reduce medication errors. *Healthcare Informatics* 2006. 23:38–41.

67. Sakowski, J., Leonard, T., Colburn, S., et al. Using a bar-coded medication administration system to prevent medication errors in a community hospital network. *American Journal of Health-System Pharmacy* 2005. 62:2619–2625.

68. Cochran, G. L., Jones, K. J., Brockman, J., Skinner, A., and Hicks, R. W. Errors prevented by and associated with bar-code medication administration systems. *Joint Commission Journal of Quality Patient Safety* 2007. 33:245, 293–301.

69. Ashcroft, D. M., and Cooke, J. Retrospective analysis of medication incidents reported using an online reporting system. *Pharmacy World & Science* 2006. 28:359–365.

70. Pavlin, J. A., Murdock, P., Elbert, E., et al. Conducting population behavioral health surveillance by using automated diagnostic and pharmacy data systems. *Morbidity and Mortality Weekly Report* 2004. 53(Suppl):166–172.

71. Grant, R. W., Wald, J. S., Poon, E. G., et al. Design and implementation of a Web-based patient portal linked to an ambulatory care electronic health record: Patient gateway for diabetes collaborative care. *Diabetes Technology and Therapeutics* 2006. 8:576–586.

72. Gurjar, R., Li, Q., Bugbee, D., et al. Design and implementation of a clinical rule editor for chronic disease reminders in an electronic medical record. *AMIA Annual Symposium Proceedings* 2006:936.

73. Heymann, A. D., Chodick, G., Halkin, H., et al. The implementation of managed care for diabetes using medical informatics in a large preferred provider organization. *Diabetes Research and Clinical Practice* 2006. 71:290–298.

74. Sanghera, N., Chan, P. Y., Khaki, Z. F., et al. Interventions of hospital pharmacists in improving drug therapy in children: A systematic literature review. *Drug Safety* 2006. 29:1031–1047.

75. Kwan, Y., Fernandes, O. A., Nagge, J. J., et al. Pharmacist medication assessments in a surgical preadmission clinic. *Archives of Internal Medicine* 2007. 167:1034–1040.

76. Hayes, B. D., Donovan, J. L., Smith, B. S., and Hartman, C. A. Pharmacist-conducted medication reconciliation in an emergency department. *American Journal of Health-System Pharmacy* 2007. 64:1720–1723.

77. Fertleman, M., Barnett, N., and Patel, T. Improving medication management for patients: The effect of a pharmacist on postadmission ward rounds. *Quality Safety Health Care* 2005. 14:207–211.

78. Carter, M. K., Allin, D. M., Scott, L. A., and Grauer, D. Pharmacist-acquired medication histories in a university hospital emergency department. *American Journal of Health-System Pharmacy* 2006. 63:2500–2503.

79. Buurma, H., De Smet, P. A., Leufkens, H. G., and Egberts, A. C. Evaluation of the clinical value of pharmacists' modifications of prescription errors. *British Journal of Clinical Pharmacology* 2004. 58:503–511.

80. Schnipper, J. L., Kirwin, J. L., Cotugno, M. C., et al. Role of pharmacist counseling in preventing adverse drug events after hospitalization. *Archives of Internal Medicine* 2006. 166:565–571.

81. Walton, S. M. The pharmacist shortage and medication errors: Issues and evidence. *Journal of Medical Systems* 2004. 28:63–69.

82. Schneider, P. J. Opportunities for pharmacy. *American Journal of Health-System Pharmacy* 2007. 64:S10–S16.

83. Woodruff, A. E., and Hunt, C. A. Involvement in medical informatics may enable pharmacists to expand their consultation potential and improve the quality of healthcare. *Annals of Pharmacotherapy* 1992. 26:100–104.

84. Wang, J. K., Herzog, N. S., Kaushal, R., Park, C., Mochizuki, C., and Weingarten, S. R. Prevention of pediatric medication errors by hospital pharmacists and the potential benefit of computerized physician order entry. *Pediatrics* 2007. 119:e77–e85.

CHAPTER 11

Tertiary Information Sources for Professionals and Patients

Philip O. Anderson and Susan M. McGuinness

CONTENTS

11.1 INTRODUCTION

When a pharmacist receives a question that requires the use of a computerized database, it is likely that he or she will use a tertiary drug information database. This chapter will review some of the major tertiary databases.

11.2 SYSTEMATIC APPROACH

No database will provide an accurate, patient-specific answer to a query if the user does not thoroughly understand the question, why it is asked, and how the information will be used. Therefore, before answering any question about a medication, it is important to gather all relevant information from the person asking the question.

A systematic approach to answering drug information questions is essential and has been well described.[1,2] The pharmacist should become familiar with this approach before answering patient-specific questions because a seemingly straightforward question can actually mean something quite different from what it initially seems to mean. The first two steps in the process are outlined next (modified from Kirkwood and Kier[2]):

- secure demographics of requestor:

    - name,

    - telephone or pager number,

    - institution or practice site if a healthcare professional, and

    - title, profession or occupation, and rank;

- obtain background information:

    - What resources has the requestor already consulted?

    - Is the request patient specific or academic?

    - What are the patient's diagnosis, other medications, and pertinent medical information?

    - How urgent is the request?

11.2.1 Secondary Sources

Once these necessary pieces of information are collected, the pharmacist should formulate a search strategy. Should PubMed or another secondary database be consulted or should a tertiary database be used? Novices often begin searching directly in a secondary database, thinking that the information there is more relevant and has a greater degree of authenticity than that in a tertiary database. However, secondary sources have several disadvantages for answering drug information questions:

- *Too much information.* The amount of information can be overwhelming, often resulting in the searcher using only a select subset of the search results, usually the most recent.

- *Poor search strategy.* Conversely, a poor search strategy can lead to too few results and missing important published information in the database.

- *Older information may be better.* The best information to use for answering the question might have been published several years ago.

- *Time consuming.* Because of the preceding issues, searching secondary sources can be time consuming. This time may not be available in the busy pharmacy.

- *Lack of context.* Information in secondary databases might not have been placed into the larger context of the disease states of interest.

- *Incompleteness.* Articles located in secondary databases are often incomplete sources that do not fully discuss benefits, side effects, and contraindications of therapy.

11.2.2 Tertiary Sources

Tertiary drug information databases overcome most of these problems. The authors of these databases have usually already performed a thorough search of secondary databases. The quality of the evidence has been assessed; the information has been evaluated for relevance and placed into context. Ideally, the author has some clinical expertise in the disease state for which the drug is used. Information from the FDA-approved labeling is usually included and supplemented with information from published literature for completeness.

Nevertheless, tertiary sources have some potential limitations, too. The user of tertiary drug information database should also consider the following:

- *Updating may be too infrequent.* Even if database updates are sent frequently (e.g., quarterly), only selected portions of the database are updated, rather than the entire database. Individual records should be consulted for the most recent update. Databases that reside on the producer's server can be updated daily and are preferable to those that reside on a local server and are updated at longer intervals (e.g., monthly, quarterly) via CD ROM, DVD, or tape.

- *Information can be overly dependent on the package insert.* A good database should include such FDA-approved information, but also supplement it with literature sources. Once drugs become generic, manufacturers usually do not spend money to investigate the drug further (e.g., new indications), so the database authors should provide this information from the primary literature. Ideally, a tertiary database lists the references used and has hot links to PubMed records of these articles for easy access.

- *Details of studies are often not given.* It may not be clear how the general information in the database applies to patients who are outliers (e.g., morbidly obese patients). The database should provide references to the primary literature so that the user can easily locate the studies and find such details, if necessary.

- *Information cannot automatically be applied to the specific patient in question.* Database users always must determine the most appropriate way to supply and apply the information as described in the systematic approach to answering questions.

Often, the best strategy for searching for drug information is first to consult a tertiary database where the needed information might be located and only then proceed to searching secondary sources, if necessary. Secondary sources can add more recent or more detailed information to that found in the tertiary database if needed. This chapter will describe and contrast some of the major tertiary drug information databases.

11.3 DRUG INFORMATION DATABASES FOR PROFESSIONALS

Four popular stand-alone drug information databases are comprehensive in scope. The main features of these databases are reviewed next. Each database also contains a variety of ancillary databases, such as FDA warnings, manufacturer contact information, normal laboratory values, and dosage calculators; however, these will not be reviewed.

11.3.1 Clinical Pharmacology

Founded in 1993 by Gold Standard (http://www.goldstandard.com), Clinical Pharmacology is one of the newer comprehensive drug information databases. The database has a clear, simple user interface. Records are divided into 11–12 subsections so that the user can reach information quickly. Drug records contain standard package insert recommendations supplemented by literature sources that can be found by clicking on the reference numbers; many of the references have hot links to PubMed.

Because of excellent consistency and deployment of controlled vocabulary throughout the database, it can be searched by several different factors, such as indication, adverse reactions, contraindications, etc. The consistent terminology also allows for a unique customizable report feature. For example, a question such as, "Which of the patient's five drugs could be causing thrombocytopenia?" can be searched simply from one screen. Drug interaction and intravenous (IV) compatibility checkers, a tablet identification feature, and patient information leaflets are included in the package as well as several preprogrammed comparison charts. The Natural Medicines Comprehensive Database is also linked to Clinical Pharmacology, providing the user with an extensive database on natural products and nutraceuticals.

11.3.2 Facts and Comparisons

Facts and Comparisons (http://factsandcomparisons.com/) is one of the oldest print drug information sources. It was originally developed as an alternative to the manufacturer-focused *Physician's Desk Reference*. Information in the main drug information database is primarily reformatted information from the package insert with some literature referencing. Other databases (e.g., Drug Interactions Facts) are more literature based and more completely referenced. The online version is called Facts and Comparisons 4.0. It has a fairly simple user interface with two search sections: one for drug information and one for disease and symptom searching. Several different databases can be searched for information, including the main database, an abbreviated database (A to Z drug facts), the drug–drug and drug–herbal interactions databases, patient information, the Review of Natural Products, nonprescription products, tablet identification, and off-label indications.

Although exclusive to Facts and Comparisons, some of these databases, such as Drug Interactions Facts and the Review of Natural Products, are produced by outside consultants.

This occasionally leads to contradictions between databases—for example, different ratings of herbal drug interactions between the two interactions databases. A strong point of the print version of Facts and Comparisons has always been its comparison charts, which are helpful in comparing drug products with similar uses and ingredients. The online version continues this tradition. Facts and Comparisons 4.0 does not contain an IV compatibility checking program, but does have a number of useful tables, such as look-alike/sound-alike drugs and don't crush/chew lists.

11.3.3 Lexi-Comp

LexiComp (http://www.lexi.com/) is one of the newer drug information databases and is available as online, desktop, and PDA (personal digital assistant) versions. Numerous databases—some produced by outside sources and some aimed at other professions such as dentists and nurses—are available in various combination packages. Information in the databases tends to be concise and sparsely referenced; this serves it well in independent tests of the PDA versions, but not in comparative studies of the more complete databases. A strong point is its pediatric database, which is expertly written and well referenced.

11.3.4 Micromedex

Micromedex (http://www.micromedex.com) was one of the first electronic database providers. Micromedex provides a number of separate databases produced by different providers that can be purchased in many combinations. Available databases are searched via a single search box and then the user selects which database to examine to find the desired information. DRUGDEX is the comprehensive general drug information database produced by Micromedex. The user can reach DRUGDEX monographs and then can either scroll through the record or jump to one of the subsections. Information is laid out in an outline format rather than in continuous paragraphs. The outline elements include information from the package insert and primary literature abstracts.

The details from the literature can be useful, but the lack of flow and integration of the outline format can be confusing. The records are thoroughly referenced, and references are provided alphabetically at the end of each drug record. However, the user must jump to the end of the record and search to locate the reference by author name and publication year. Other database features include drug interaction and IV compatibility checkers and the ability to compare two drugs at a time in an abbreviated (DrugPoints) format. Micromedex also has some databases from other producers such as Martindale from Great Britain, which offers foreign drug information, and four databases on the reproductive effects of drugs and chemicals.

11.3.5 Embedded Databases

In addition to these stand-alone references, most pharmacy computer systems have embedded databases that automatically perform functions such as drug interactions and allergy checking. Such programs are available from providers such as First Data Bank (the most widely used) and Medi-Span. The programs may also have drug information that can be accessed by clicking on links in the program. These sources have not been compared as rigorously as have the stand-alone programs, except for their drug–drug interaction checking modules (discussed later).

11.4 COMPARISONS OF DATABASES

11.4.1 Overall Comparisons

In addition to the various features of the databases and their user interfaces, the information provided often differs. Several studies have been published that directly compared the quality of data and their ability to answer routine drug information inquiries. In evaluating these studies, one must remember that they are a snapshot in time and do not necessarily represent the databases today because changes (presumably improvements) and updates are continually being made. Another factor in any study is the perceptions and needs of the rater. Some raters might rate comprehensiveness as a more important feature than accessibility or ease of use, while others might have different priorities. The more recently published studies are summarized next.

A study in 2004 illustrated user preferences well.[3] Four pharmacy students, four pharmacy faculty, and four medical librarians used five electronic drug information sources to answer 10 drug information questions. Pharmacy students and faculty rated e-Facts (now Facts and Comparisons 4.0) and Lexi-Drugs highest because of their conciseness; medical librarians ranked DRUGDEX highest because of its completeness, layout, and references supporting the data. The electronic *Physicians Desk Reference* and AHFS Drug Information were ranked lowest by all.

A more comprehensive study published in 2007 compared more electronic drug information sources.[4] The authors used 158 drug information questions from 15 categories of question types to evaluate five subscription and three free online databases. Databases were evaluated on their ability to answer the question, the comprehensiveness of the answer, and the ease of use. These scores were combined into an overall composite score for each database and the following rankings were determined: (1) Clinical Pharmacology, (2) Micromedex (DRUGDEX and Identidex), (3) Lexi-Comp online, and (4) Facts & Comparisons 4.0. RxList and Epocrates premium formed a second tier and the free version of Epocrates performed markedly worse than the other databases. The paid databases performed better than the free databases.

When the same question set was used to evaluate the PDA versions of these databases, the rankings changed somewhat.[5] Lexi-Drugs performed best, Clinical Pharmacology On-Hand was second, Epocrates Rx Pro was third, mobileMicromedex (now Thomson Clinical Xpert) was fourth, and the free version of Epocrates Rx was again last.

This study illustrates an important point. Databases designed for the PDA (Lexi-Drugs, Epocrates Rx Pro) tend to do better on PDAs than "stripped down" versions of larger databases, whereas computer-based databases (e.g., Clinical Pharmacology, Micromedex) outperform PDA databases when compared on full-sized computers.

11.4.2 Drug Identification

A now rather dated study has been the only one to compare the ability of databases to identify solid oral dosage forms by their markings in 2001 and 2002.[6] Using seven online databases, an attempt was made to identify 500 solid oral dosage forms brought to the hospital by patients. Identidex (from Micromedex) and Ident-A-Drug (from Therapeutic Research [publishers of *The Pharmacist's Letter*]) performed best, both at 86% of dosage forms identified. RxList was third at 71%; Clinical Pharmacology On-Hand and Lexi-Comp both identified about 68% of dosage forms. The PDR and Facts and Comparisons were far behind at 43 and 25%, respectively. The likelihood of identifying solid oral dosage forms was better if two databases were used. It seems that improvements have been made in this function of some databases (e.g., Clinical Pharmacology, Facts & Comparisons 4.0), so current results would probably differ.

11.4.3 Drug Interactions

Most of the major electronic drug information databases have a drug–drug interaction checking feature. On no other topic do tertiary reference sources differ more than that of drug interaction checking. Several studies have found marked differences among databases in their ability to detect and rate drug–drug interactions.

An early study comparing various databases found that the number of drug interactions classified as "major" varied markedly by database.[7] Drug Reax (Micromedex) classified 1,841 reactions as "major" and Drug Interactions Facts (Fact & Comparisons 4.0) listed 225. Drug Interactions: Analysis and Management (Hansten and Horn) listed 200, and Evaluation of Drug Interactions (APhA) listed only 88. Only nine (2.2%) of the major drug interactions were listed in all four compendia! Possible reasons for discrepancies include differences in the criteria and judgments as to what constitutes a "major" interaction, extrapolation of interactions of one drug to others in the class, and differences in primary information sources that databases use (e.g., foreign language sources, unpublished manufacturer information). In addition, although some of the databases have three severity levels, others have four or five.

In 2003, Australian investigators compared 1,095 drug interactions of 50 drugs listed in at least one of four drug interaction sources: the British National Formulary; the French source, Vidal; Drug-Reax; and Drug Interactions Facts.[8] Great discrepancies were noted between sources. Between 14 and 44% of "major" reactions listed in any one source were not even listed as interactions in other sources.

These discrepancies among databases are not limited to tertiary "look-up" sources. A series of articles using a standardized set of patient cases found that drug interaction checking by pharmacy computer systems performed poorly. In the initial study,[9] nine different computer programs deployed among 516 community pharmacies in Washington state failed to detect clinically relevant drug interactions one-third of the time. The number of "false-positive" drug interaction reports varied, also. Interestingly, results varied considerably even between pharmacies using the same database, indicating that interface settings that can be changed by the user affect performance.

In 2004, virtually the same set of cases was used to test the system of eight community and five hospital pharmacies in the Tucson, Arizona, area.[10] Community pharmacy systems performed better than in the previous hospital study, detecting 88% of important interactions, whereas hospital systems detected only 38%. When the same set of patient cases was later applied to PDA software in 2004, the results were better than with the large pharmacy systems.[11] The PDA programs detected between 81 and 100% of interactions, with an average of 95%.

In Switzerland, nine drug–drug interactions sources were screened on several preliminary criteria. Programs failing the initial screening included the British National Formulary, Epocrates MultiCheck, Stockley's Drug Interactions, the Medical Letter, and Vidal. Only four passed the initial selection: Drug Interactions Facts, Drug-Reax (Micromedex), Lexi-Interact, and Pharmavista, a German-language program.

These four programs were then subjected to further testing using simulated questions. Drug Interactions Facts was found to be the least comprehensive and Lexi-Interact the most. Of the three databases used in the United States, Drug Interactions Facts had the highest specificity and lowest sensitivity, whereas Lexi-Interact had the lowest specificity and highest sensitivity. Drug-Reax was intermediate on both scores. In 16 actual patient profiles, all programs had high negative predictive values (no reaction reported when none exists). Positive prediction value (ability to detect an actual interaction) ranged from about 60 to 70%, except for Lexi-Interact, which was lower at 36%.[12]

In short, no single drug–drug interaction database can currently be relied upon for drug interaction checking. Use of more than one source of drug interaction information is advisable. For example, a source with broad inclusion criteria partially based on FDA-approved drug labeling that may have theoretical interactions can be compared with another source that has clear criteria for inclusion and good literature documentation to support its conclusions. Finally, the user must employ good clinical judgment to apply drug–drug interaction data to a specific patient.

11.4.4 Herbals and Nutraceuticals

It is not entirely unexpected to find differences among databases in information on "complementary" therapies because the underlying primary literature on these is often poor or completely absent. Two relatively recent studies compared electronic databases' and published books' abilities to answer questions posed to drug information services regarding these products.[13,14] Both found that the Natural Medicines Comprehensive Database (from Therapeutic Research as a book and database and now incorporated into Clinical Pharmacology) performed best, although it was far from complete. In one study it answered 61% of questions completely and 16% partially. In the other study, it was very helpful in 62% of questions and somewhat helpful in 20% of questions. AltMedDex (Micromedex) ranked second in both studies, and the Review of Natural Products (Facts & Comparisons) and the Herbal Companion to AHFS DI ranked somewhat lower. Both studies found the PDR for Herbal Medicines to perform very poorly.

In another study, four natural medicine databases were compared in their ability to answer 102 questions on complementary therapies in 10 categories (e.g., dosage, pharmacokinetics, interactions). Databases were rated on scope (presence or absence of an answer), completeness, and ease of use; in addition, they were given a composite score of the three domains. For completeness, the ranking was, from best to worst, Natural Medicines Comprehensive Database, the Natural Standard, AltMedDex, and Lexi-Natural.

An overall composite score ranked databases in the same order. All online databases were about the same in their ease-of-use scores. The study also compared PDA versions to the full databases and found that both AltMedDex and the Natural Standard online databases performed better than their respective PDA versions.[15]

11.4.5 Infectious Disease Agents

The ability of 14 databases (eight subscription and six free) to answer questions related to infectious diseases was studied. Sixteen categories of questions (e.g., dosage, pharmacokinetics, interactions) were studied across five groups of agents (e.g., antivirals, antibacterials) for a total of 147 questions. The databases fell into three statistically different categories when ranked on scope; the best were Micromedex, Medscape Drug Reference, Lexi-Comp-AHFS, AHFS, and Clinical Pharmacology. Three categories were found for completeness; the best were Micromedex, DailyMed, Internet Drug Index, Medscape Drug Reference, AHFS, Clinical Pharmacology, Facts & Comparisons Online, Lexi-Comp-AHFS, and DIOne. The only databases found to have no errors in this study were Micromedex, AHFS, and Medscape Drug Reference. PEPID PDC had statistically more errors (seven) than the other databases. Somewhat surprisingly, the Johns Hopkins Antibiotic Guide had two errors in dosages out of 15 questions in this category.[16]

11.5 DRUG INFORMATION DATABASES FOR PATIENTS

Healthcare consumers obtain information about their medications from a variety of sources other than their healthcare providers, but usually do not have access to the commercial databases discussed previously. Drug advertising pervades all communications media, and the Internet provides a wealth of information, much of which is trustworthy but some of which is not. The free availability of drug information helps patients to participate more actively in decisions about their care, and it may also cause them to ask more questions about their drug therapy. Patients may ask providers for medications they see in advertisements without really understanding the benefits and risks of taking those medications. They may fail to comply with recommended drug therapy because they hear of "rare but serious" side effects (e.g., in television ads), or they may misunderstand the reason a drug was prescribed because they read about it on a Web site that was not authoritative or relevant to their disease.

As increasing numbers of consumers acquire computers and Internet connections in their homes, healthcare professionals are encountering increasing numbers of patients who are informed about their conditions and medications through the Internet. The Pew Internet and American Life Project reports that 80% of Internet users search for health-related issues and that "15% of health seekers say they 'always' check the source and date of the health information they find online, while another 10% say they do so 'most of the time.' Fully three-quarters of health seekers say they check the source and date 'only sometimes,' 'hardly ever,' or 'never.'"[17]

If consumers use information to make decisions about their healthcare, those who trust bad information are at risk. Most consumers trust information found on the Internet, but their confidence does not correlate with accuracy of the information.[18] Pharmacists today serve patients who are more informed (or misinformed) than ever and have more questions about their healthcare. They must be aware of sources of patient information and help guide patients to trustworthy sources.

Many people can evaluate a Web site's validity fairly quickly; for example, they may accept information coming from a government agency and automatically reject sites from a commercial domain. Search engines such as Google can usually be trusted to return reliable sites in the first page of search results, with more questionable sites farther down the list. An intuitive sense of what is trustworthy information is often all that is needed to find good information, but it is helpful for pharmacists to articulate specific evaluation criteria in order to help patients locate and use information on the Web.[19] Many professional organizations provide guidelines for evaluating health information on the Web. The Health on the Net Foundation (HON) is a nonprofit organization that developed a code of ethics for health Web sites and offers certification for Web sites that adhere to their "HONcode."[20] The U.S. National Cancer Institute also provides guidelines.[21]

11.5.1 Guidelines

The sources cited earlier provide many guidelines for evaluating health information on the Web, including the following criteria:

- *Authority.* Authors of Web sites should be easily identifiable, and their credentials should be listed. The domain can also provide a clue to authority; sites from the .gov and .edu domains are often more reliable than .com sites. Search engines have tools for limiting searches to specific domains.

- *Accuracy.* The information is supported by references to research and other authoritative Web sites. Look for factual or grammatical errors in the content.

- *Purpose.* If the site is intended for advertising, it should be identified as such.

- *Currency.* The site should show when it was last updated and should be well maintained. Dead links could be a clue that the site is not consistently updated.

- *Policy statements.* Look for editorial policy statements explaining the sources of information and policies for using the information. Good health Web sites should always state that the information is not intended to replace the advice of healthcare providers.

- *Indicators of quality.* Look for certifications such as the HONcode. This will indicate that the site is trustworthy because it adheres to the HONcode of ethics for presenting health information on the Web. But keep in mind that the absence of a HON certificate does not necessarily indicate that the site is untrustworthy.

11.5.2 Health Literacy

These guidelines should be shared with patients who seek health information on the Internet. Pharmacists assisting patients with health information must also consider patients' varied levels of health literacy. The Institute of Medicine (IOM) defines health literacy as "the degree to which individuals have the capacity to obtain, process, and understand basic information and services needed to make appropriate decisions regarding their health."[22] Health literacy involves the ability of consumers to communicate effectively with healthcare providers and follow their recommendations. The IOM reported that almost half of adult Americans lack the skills necessary to understand health information. Low health literacy is associated with poor outcomes in the areas of disease management and medication compliance.

Education, ethnicity, and socioeconomic status may affect health literacy, but age is the factor most strongly associated with low health literacy.[23] This presents a huge challenge because age is also associated with chronic conditions and increased medication use. Pharmacists have an important role in medication therapy management, so they must be able to assess health literacy and help patients with low health literacy. A study of medication use after discharge from a hospital showed that over 25% of patients failed to remember the names, doses, or purposes of newly prescribed medications and only 11% could recall being told of any adverse effects.[24] A number of tests assess health literacy[25,26]; however, it may not always be possible to administer such tests in the pharmacy practice setting. Pharmacists can help patients by being sensitive to indicators of low health literacy and providing information at a level that patients can understand.[27–29]

Health literacy should not be confused with literacy in reading and writing. Many intelligent, educated people have low health literacy. Pharmacists must be aware that most patients need some help to understand information about their medications.

11.6 RESOURCES FOR PATIENTS

In addition to helping patients evaluate health information, understanding their health literacy, and tailoring information to their needs, pharmacists should be aware of high-quality tertiary sources on the Web that are intended for patients. Although search engines often retrieve high-quality Web sites in the first page of results, they often miss information contained in large megasites or portals. It is therefore important to know about these Web resources for consumers.

11.6.1 MedlinePlus

Many subscription-based resources, such as Clinical Pharmacology and Micromedex, include drug information for patients, but the best starting point for free consumer health information on the Web is medlineplus.gov, which provides authoritative, noncommercial information that is well organized and available in English and Spanish. MedlinePlus is easy to use. Information on drugs, herbals, and supplements can be found by entering the generic or trade name into a search form or by browsing alphabetical lists of drug names. Each entry, or monograph, is an overview of the drug's indications, doses, adverse effects, and instructions on use. The monographs on herbals and supplements include information about the quality of the documentation supporting the use of the product for a variety of indications. These monographs serve as patient handouts and are freely available anywhere there is an Internet connection.

MedlinePlus is compiled by the National Library of Medicine (NLM), National Institutes of Health, and other government agencies and health organizations. The site is updated daily and individual health topics are reviewed at least every 6 months. It includes an illustrated medical encyclopedia, directories for finding doctors or hospitals, health news, tutorials and videos, and links to many other excellent information sources. Drug information is provided by the American Society of Health-System Pharmacists and herbal and supplement information by Natural Standard. MedlinePlus resources are written in lay language, but also provide links to primary literature. The NLM provides a drug information portal (http://druginfo.nlm.nih.gov/drugportal/drugportal.jsp) that links searchers to drug information from NLM-sponsored Web sites, including MedlinePlus, LactMed (a database on drug use during breastfeeding), DailyMed (a listing of FDA-approved package inserts), clinical trials, and more.

11.6.2 NCCAM

Another government-sponsored Web site that is especially useful in pharmacy is the National Center for Complementary and Alternative Medicine (NCCAM; http://nccam.nih.gov/). NCCAM provides general information about safe and effective uses of CAM and links to monographs about herbal therapies and other alternative treatments. The monographs usually include a section on "what the science says," which describes in lay language the effectiveness of the therapy. NCCAM monographs usually include links to MedlinePlus, where users can find more detailed discussions of the quality of the evidence.

11.6.3 FDA Consumer Drug Information

The FDA site (http://www.fda.gov/cder/drug/DrugSafety/DrugIndex.htm) lists drugs by name. Each drug monograph includes a patient information sheet that explains what the drug is for, how it works, precautions, and how the drug should be taken. Information for patients on FDA-issued warnings about the drug is also included.

11.6.4 CDC

The Centers for Disease Control and Prevention (http://www.cdc.gov/) is an invaluable source of information about many health topics. It is especially useful to both consumers and health professionals as a source of information on immunizations.

11.7 CONCLUSION

Tertiary databases are extremely useful tools for health professionals. However, they cannot be considered "generically equivalent." Each database has strengths and weaknesses that can affect patient care. With the possible exception of drug interaction checking, computer-based programs usually outperform PDA programs when they are compared head to head. The pharmacist not only must have good searching skills to use the databases successfully, but also must have good drug information skills to interpret, apply, and supply the information contained in the databases. It is incumbent on the users of these databases to use all aspects of their pharmacy education to make the best use of them.

Numerous authoritative, free Web resources are available for the lay public on drugs, herbals, nutraceuticals, and other health matters. By directing patients to these reliable sites, pharmacists can have confidence that the information will be accurate and up to date.

REFERENCES

1. Calis, K. A., and Sheehan, A. H. Formulating effective responses and recommendations: A structured approach. In *Drug information. A guide for pharmacists,* 3rd ed., eds. P. M. Malone, K. L. Kier, and J. E. Stanovich, 39–59. New York: McGraw–Hill, 2006.
2. Kirkwood, K. F., and Kier, K. L. Modified systematic approach to answering questions. In *Drug information. A guide for pharmacists,* 3rd ed., eds. P. M. Malone, K. L. Kier, and J. E. Stanovich, 29–37. New York: McGraw–Hill, 2006.
3. Kupferberg, N., and Jones Hartel, L. Evaluation of five full-text drug databases by pharmacy students, faculty, and librarians: Do the groups agree? *Journal of the Medical Library Association* 2004. 92:66–71.
4. Clauson, K. A., Marsh, W. A., Polen, H. H. Seamon, M. J., and Ortiz, B. I. Clinical decision support tools: Analysis of online drug information databases. *BMC Medical Informatics Decision Making* 2007. 7:7.
5. Clauson, K. A., Polen, H. H., and Marsh, W. A. Clinical decision support tools: Performance of personal digital assistant versus online drug information databases. *Pharmacotherapy* 2007. 27:1651–1658.
6. Raschke, C. G., Hatton, R. C., Weaver, S. J., and Belgado, B. S. Evaluation of electronic databases used to identify solid oral dosage forms. *American Journal of Health-System Pharmacy* 2003. 60:1735–1740.
7. Abarca, J., Malone, D. C., Armstrong, E. P., et al. Concordance of severity ratings provided in four drug interaction compendia. *Journal of the American Pharmacists Association* 2004. 44:136–141.
8. Vitry, A. I. Comparative assessment of four drug interaction compendia. *British Journal of Clinical Pharmacology* 2007. 63:709–714.
9. Hazlet, T. K., Lee, T. A., Hansten, P. D., and Horn, J. R. Performance of community pharmacy drug interaction software. *Journal of the American Pharmaceutical Association* (Wash) 2001. 41:200–204.
10. Abarca, J., Colon, L. R., Wang, V. S., et al. Evaluation of the performance of drug–drug interaction screening software in community and hospital pharmacies. *Journal of Managed Care Pharmacy* 2006. 12:383–389.
11. Perkins, N. A., Murphy, J. E., Malone, D. C., and Armstrong, E. P. Performance of drug–drug interaction software for personal digital assistants. *Annals of Pharmacotherapy* 2006. 40:850–855.
12. Vonbach, P., Dubied, A., Krahenbuhl, S., and Beer, J. H. Evaluation of frequently used drug interaction screening programs. *Pharmacy World & Science* 2008. 30:367–374.
13. Sweet, B. V., Gay, W. E., Leady, M. A., and Stumpf, J. L. Usefulness of herbal and dietary supplement references. *Annals of Pharmacotherapy* 2003. 37:494–499.
14. Walker, J. B. Evaluation of the ability of seven herbal resources to answer questions about herbal products asked in drug information centers. *Pharmacotherapy* 2002. 22:1611–1615.
15. Clauson, K. A., Peak, A. S., Marsh, W. A., DiScala, S., and Bellinger, R. R. Clinical decision support tools: Focus on dietary supplement databases. *Alternative Therapies in Health and Medicine* 2008. 14:36–40.
16. Polen, H. H., Zapantis, A., Clauson, K. A., Jebrock, J., and Paris, M. Ability of online drug databases to assist in clinical decision-making with infectious disease therapies. *BMC Infectious Diseases* 2008. 8:153.
17. Fox, S. *Online health search 2006.* Washington, D.C.: Pew Internet & American Life Project.

18. Lau, A. Y., and Coiera, E. W. Impact of Web searching and social feedback on consumer decision making: A prospective online experiment. *Journal of Medical Internet Research* 2008. 10:e2.

19. Hunter, T. S., ed. *e-Pharmacy: A guide to the Internet care zone.* Washington, D.C.: American Pharmaceutical Association, 2002.

20. Health on the Net Foundation (available at http://www.hon.ch/HONcode/Pro/Conduct.html). Geneva, Switzerland: HON (accessed May 9, 2009).

21. National Cancer Institute. 2009 (available at http://www.cancer.gov/cancertopics/factsheet/information/internet). Bethesda, MD: National Institutes of Health (accessed May 9, 2009).

22. Nielsen-Bohlman, L., Panzer, A. M., and Kindig, D. A., eds. *Health literacy: A prescription to end confusion.* Committee on Health Literacy, Board on Neuroscience and Behavioral Health (Institute of Medicine). Washington, D.C.: National Academies Press, 2004.

23. Keller, D. L., Wright, J., and Pace, H. A. Impact of health literacy on health outcomes in ambulatory care patients: A systematic review. *Annals of Pharmacotherapy* 2008. 42:1272–1281.

24. Maniaci, M. J., Heckman, M. G., and Dawson, N. L. Functional health literacy and understanding of medications at discharge. *Mayo Clinic Proceedings* 2008. 83:554–558.

25. Davis, T. C., Crouch, M. A., Long, S. W., et al. Rapid assessment of literacy levels of adult primary care patients. *Family Medicine* 1991. 23:433–435.

26. Parker, R. M., Baker, D. W., Williams, M. V., and Nurss, J. R. The test of functional health literacy in adults: A new instrument for measuring patients' literacy skills. *Journal of General Internal Medicine* 1995. 10:537–541.

27. Brown, L., Upchurch, G., and Frank, S. K. Low health literacy: What pharmacists can do to help. *Journal of the American Pharmacists Association* 2006. 46:4–11.

28. Safeer, R. S., Cooke, C. E., and Robertson, T. A. Pharmacy measures to improve medication use through health-literacy principles. *Managed Care Interface* 2007. 20:37–41.

29. Tkacz, V. L., Metzger, A., and Pruchnicki, M. C. Health literacy in pharmacy. *American Journal of Health-System Pharmacy* 2008. 65:974–981.

CHAPTER 12

PDAs and Handheld Devices

Joseph J. Ennesser

CONTENTS

12.1 INTRODUCTION

What is a PDA? If this question had been asked just a few years ago, the answer would have been "a Palm Pilot or Pocket PC," but today the term PDA fits a broader spectrum of devices. Technically (and specifically) speaking, PDA is short for "personal digital assistant," which is a handheld electronic device that allows the user to store, quickly access, and edit personal information (contacts, calendar, etc.). This information can then be synchronized with a computer to back up the data. In their simplest form, PDAs are well identified by this definition.

However, as handheld devices have continued to change rapidly in both form and function, application of the term PDA has broadened. There are several types of handheld information devices: mobile phones, digital cameras, personal media players, data storage devices, and mobile Internet devices; however, increasingly, one single device incorporates several or all of these functions and true PDAs have virtually disappeared. In fact, with the introduction of the Apple iPhone in 2007, a new revolution in PDA development and advancement began to compete with the iPhone, which really is a portable computer that also happens to be a phone.

Handheld devices can help us organize our personal and professional lives, but they are not a replacement for practical knowledge. However, as the amount of information readily available continues to increase, they can allow us to access, interact with, and apply this information quickly to our work flow and decision-making processes. They allow us to carry compact volumes of information that are fully searchable and often editable. This is what makes handheld devices so useful to those in the health sciences, especially pharmacists. In a device not much bigger than a deck of cards, we can carry more than one full drug information reference, a dosage form identifier, a medical dictionary, an infectious disease reference, medical calculators, and much more. First-time PDA users may feel that it is quicker and easier to look up information from a paper reference, but for obtaining the data most commonly referenced in daily work, using a PDA can be much faster, especially when it is necessary to find several pieces of information.

This chapter will discuss handheld devices and their application to pharmacy practice. The focus will be mainly on PDAs, available drug references, and other programs useful to pharmacists. Examples are given on how PDAs can help a user answer questions and other practical applications in healthcare.

12.2 PERSONAL DIGITAL ASSISTANTS

Although many people refer to PDAs generically as "Palm Pilots," this is technically incorrect. The "Pilot" was an early model PDA made by Palm Computing, Inc. in 1996. Interestingly, Palm was not the first to make a PDA; Apple Computer, Inc. produced a PDA from 1993 to 1998 called the Apple Newton. However, due to its high price, large size, and handwriting recognition problems, the product was not successful. When Palm introduced its Pilot in 1996, the device's small size, easy data synchronization, and simple handwriting input method made it an instant success and the name stuck with consumers.

In their most basic form, all PDAs share some characteristics: storage of personal data (calendar, contacts, "to do" lists), touch screens, solid-state memory, and the ability to be powered on instantly. For nearly a decade, there were only two types of PDAs: Palm and Pocket PC, which were differentiated by their operating systems (OS). However, there are now two additional major types: the BlackBerry and the Apple iPhone (note that any reference to the iPhone also generally applies to the Apple iPod Touch device). A few other platforms do exist, such as the Google Android operating system, but the current small number of available applications makes them of limited use, especially in healthcare. Until recently, BlackBerry devices did not have a touch screen interface, resulting in fewer applications of interest to those in healthcare, also.

All true PDAs have a touch screen that allows the use of a stylus or a finger to interact with the device. Users navigate screens and programs by tapping on menus and icons. Text is typically entered by using a physical keyboard on the device or a virtual one on the screen. Unique to the Palm and Pocket PC, a second method of text entry called "graffiti" can be used to write one letter at a time into an area at the lower portion of the PDA's screen. Graffiti is a full alphabet, number, and symbol set of abbreviated characters that allow for input with one stylus stroke, greatly increasing input efficiency (Figure 12.1). Although it may seem slow to enter one character at a time, after some practice a user can enter strings of text rather quickly.

Recently, several mobile phone devices that look like PDAs with integrated keyboards but lacking touch screens have entered the market and are generally called "smart phones." The lack of a touch screen may prevent them from running many programs of interest to pharmacists. Those interested in purchasing a PDA should be aware of this fact when deciding on a device.

FIGURE 12.1 Palm OS Graffiti 2 character set. (Palm, Inc. 2002. Palm OS Garnet, version 5.4.0. Palm Operating System, Sunnyvale, California. Available at http://www.palm.com)

One of the greatest advantages of PDAs is their ability to go from an "off" state to a fully operational "on" state instantly—unlike personal computers, which require time to boot up. Rather than using a spinning hard disk like a typical computer, PDAs use solid-state memory. This distinction is what allows PDAs the instant "on" functionality and also extends the PDA's battery life because reading and writing data to solid-state memory use much less power than maintaining a spinning disk.

Solid-state memory comes in two main types—ROM and RAM; PDAs use both. ROM retains the stored data even after a complete loss of power; RAM requires a continuous low-level power supply to maintain the stored data. Even a momentary power loss results in RAM losing all of its stored data. RAM continuously draws power even when a PDA is turned off. Doubling the amount of RAM doubles the continuous power drain. Thus, although a device with only ROM (such as a digital camera with a flash memory card) would hold its data indefinitely, a device with only RAM would lose all its data as soon as the battery died.

Why use RAM? The answer is simple: because ROM is much slower in data read/write cycles than RAM. Using only ROM in a handheld device would make it too slow to use. Until around 2007, the majority of PDAs used ROM to store only the operating system and factory-installed files, and all the user's data and programs were stored in RAM (Figure 12.2). Managing memory in this way presents one important problem: If the PDA loses power—for even a moment—the user loses all personal data and programs stored on it. Once power is restored to the device (by charging or replacing the batteries), the user finds the device in the same state as when it was new from the factory. Only data that had been previously backed up to a computer can be recovered and any data added after the last synchronization will be lost forever.

Newer PDAs use a memory management scheme in which all files are stored in ROM and RAM is only used for actively running programs and files (Figure 12.3). In order to use

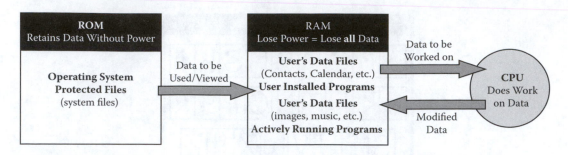

FIGURE 12.2   Classical PDA memory management scheme; lose power = lose data.

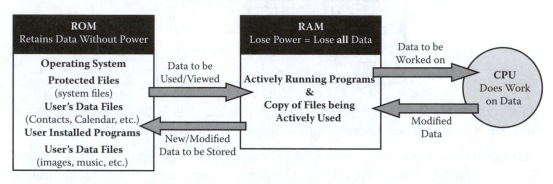

FIGURE 12.3   Current PDA memory management scheme; lose power = *no* data loss.

or view a particular program or data file, it is loaded from ROM into RAM, allowing the CPU to modify the data. Using this scheme, a complete loss of power would result in no loss of data. This is very similar to the way in which full-sized computers currently manage memory; in Figure 12.3, ROM would be replaced by the computer's hard disk. One down-side results from this change: Devices now take longer to load large data files because they must be copied from ROM to RAM before being used. Examples of large data files would be an address book or calendar with more than 1,000 entries (e.g., a business professional) or a database of medications and diseases. In time, this will likely be overcome as memory speeds continue to increase.

12.3 OPERATING SYSTEMS COMPARED

All PDA operating systems have appointment calendars, address books, "to do" lists, e-mail, and notepad programs that can synchronize their data with those on the main personal computer. All devices have a suite of additional programs installed and will allow the user to download other applications ("apps" for short). Some devices require the use of a personal computer to install apps; others do not. Currently, each of the major operating systems, with the exception of the iPhone, allow the user to download and install apps without restrictions on who developed the app or where it came from (i.e., e-mail, the Internet, etc.). In comparison, apps for the iPhone must be approved by Apple and can only be downloaded from Apple via its online store, even if the application is free.

Although Pocket PCs are not full Windows computers, users will find that the Windows Mobile OS is designed similarly to the PC Windows version. Files are stored in a directory

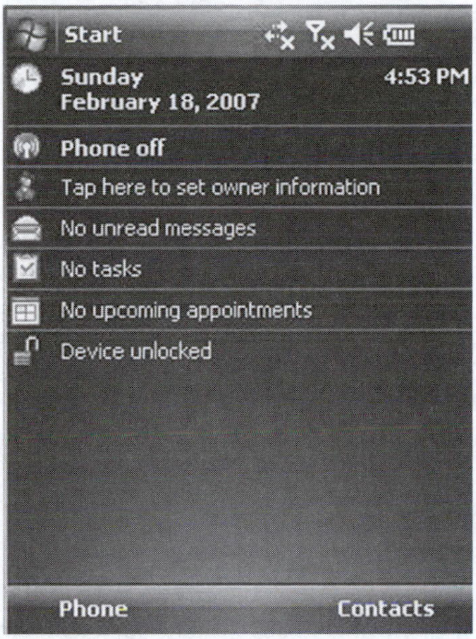

FIGURE 12.4 Windows Mobile 6 Today screen. (Microsoft, Inc. 2007. Windows Mobile Today screen, version 6, Windows Mobile Operating System, Redmond, Washington. Available at http://www.windowsmobile.com)

tree similar to Windows XP and Vista with the "My Documents," "My Music," "My Pictures," and "Programs" folders. Upon powering on a Pocket PC, users are presented with the Windows Mobile "Today Screen" (Figure 12.4). This screen is customizable and can display program icons and specific data such as upcoming calendar appointments, battery and memory status, unread e-mail notification, and other program data. Users will also find the Windows "Start" button, which is used to navigate to programs, settings, and file folders.

The Palm and iPhone operating systems are designed in a much simpler manner. Upon powering on the device, the user is presented with a screen displaying program and settings icons (Figure 12.5). There is no start button because all program and settings icons can be seen by scrolling through multiple "home" screens. No directory tree in which files are organized is readily evident; rather, these types of folders are "hidden" on the device but can be viewed using third-party software.

These three operating systems have one major difference. Palm and iPhone devices allow the user to open only one program at a time. Windows Mobile devices allow the user to have several programs open at the same time (as much as RAM will allow). The importance of simultaneously running programs can be demonstrated by a pharmacist needing reference points from a drug information database to perform a calculation requiring another program or calculator. Using Windows Mobile, the user can have the calculator and the reference program open at the same time. The same scenario on a Palm or iPhone would require closing the drug reference and losing the reference location to open the calculator. The user would then have to launch the drug reference again when done with the calculator.

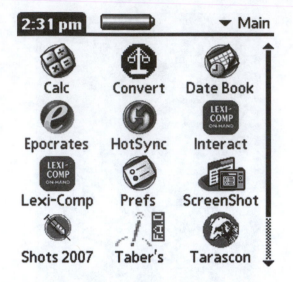

FIGURE 12.5 Palm OS main screen. (Palm, Inc. 2002. Palm OS Garnet, version 5.4.0. Palm Operating System, Sunnyvale, California. Available at http://www.palm.com)

As a consequence of being able to run multiple programs simultaneously, Windows Mobile devices tend to "freeze up" more frequently because, as programs are left open, the limited amount of RAM becomes filled, preventing further operations from occurring until these programs are closed. Interestingly, this is the reason Apple has said the iPhone will not allow simultaneous applications to run; however, there has been speculation that future versions of the iPhone OS will allow it. The next version of specifications for the Palm OS, named "Web OS," states that these devices will also allow simultaneous applications to run.

12.4 DRUG DATABASE SOFTWARE PROGRAMS FOR PDAS

Selecting pharmacy-related software for use on a PDA can be a confusing task. There are almost too many differences in database information, price, and platform availability to discuss and compare. Appendix 12.1 lists the names of some programs and their availability; comparative performance of PDA programs is discussed in Chapter 11. Most pharmacy-based programs start with a core drug database and then allow additional database or functionality upgrades at an additional charge. Typically, pricing is on an annual subscription basis.

The drug databases comprise individual drug references containing dosage, drug interactions, adverse effects and contraindications, pregnancy and breastfeeding information, and some pharmacokinetic data. The publishers can continuously update the information (typically several times per week); these updates are propagated to the PDA when the user synchronizes it with the personal computer or wirelessly if the feature is available on the PDA. With respect to PDA type, it does not matter much whether the program is being run on a Pocket PC, Palm, or an iPhone device: Once the program is launched, the user experience is nearly the same for all of the devices.

Over 10 major drug database programs are available for PDA users at the time of writing this chapter. The accuracy and completeness of the various PDA databases is reviewed in Chapter 11. Here, the focus will be on comparison of user interface and organization of information in two of the drug database programs most commonly used by pharmacists: Lexi-Drugs and Epocrates Rx.[3,4] Lexi-Drugs is a digital adaptation of the well-known paperback reference Lexi-Comp *Drug Information Handbook*; Epocrates was started in 1998 specifically as an electronic drug database for PDAs.

As evaluated at the time of this writing, Epocrates Rx is free for unlimited use and Lexi-Drugs pricing starts at $75 per year depending on PDA. Lexi does offer a free trial of 20 uses on Palm and Pocket PC devices for any of its databases, but not in the iPhone due to limitations by Apple. Both offer upgraded packages at additional cost. The Epocrates Rx database contains more than 3,300 drug names and Lexi-Drugs lists greater than 7,000. Epocrates tracks what information is viewed and the program reports this back to the publisher when updating; however, the company's Web site states that the information is used only to prioritize information updates and improve the product. Epocrates states that it does not sell or provide this information to drug companies or third parties.[5]

In terms of similarities, both programs allow the user to search for a drug name (brand and generic) by typing the first few letters of the name. Both can sort drugs alphabetically or by pharmacologic class (Figures 12.6 and 12.7). Sorting by class can be helpful in viewing drugs within a specific class for a particular condition (e.g., angiotensin II receptor blockers for hypertension).

Once a user views a specific drug monograph (by clicking on the drug name), the differences between the two programs become clear. Lexi-Drugs (Figure 12.8) presents the detailed information in a paragraph style (similar to the *Drug Information Handbook*). Epocrates (Figure 12.9) displays the information in a list format. To view more information

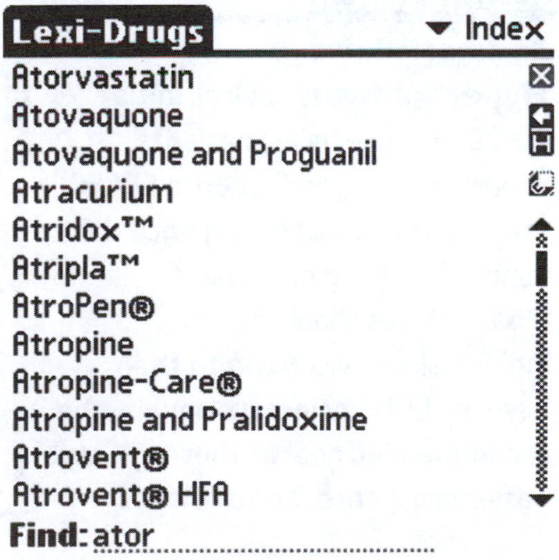

FIGURE 12.6   Lexi-Drugs main screen. (Lexi-Comp, Inc. 2007. Lexi-Drugs Essentials, file date December 28, 2007. Palm Operating System, Hudson, Ohio. Available at http://www.lexi.com)

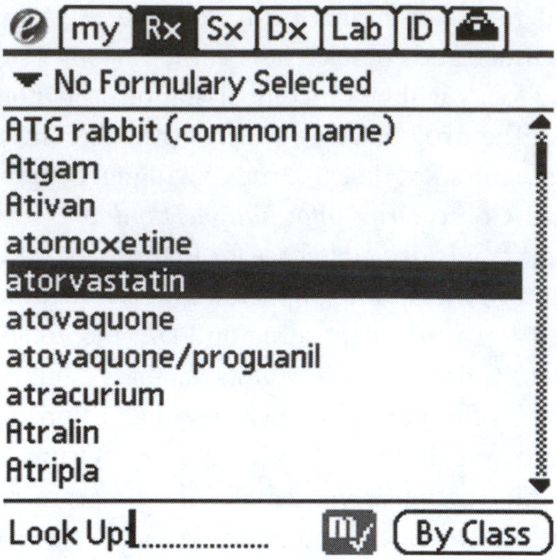

FIGURE 12.7 Epocrates Rx main screen. (Epocrates, Inc. 2007. Epocrates Rx Pro, version 8.10. Palm Operating System, San Mateo, California. Available at http://www.epocrates.com)

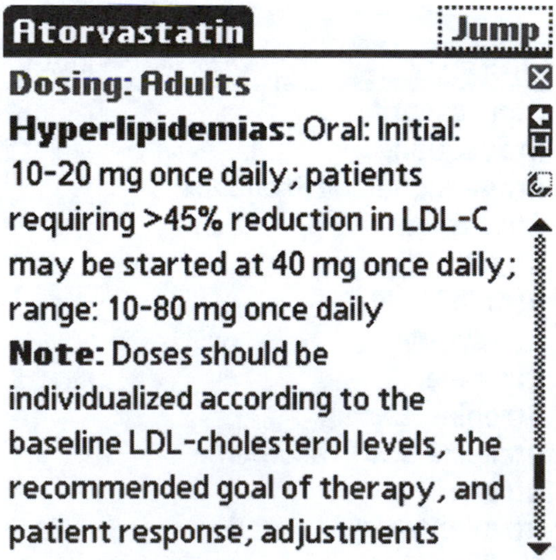

FIGURE 12.8 Lexi-Drugs atorvastatin adult dosing. (Lexi-Comp, Inc. 2007. Lexi-Drugs Essentials, file date December 28, 2007. Palm Operating System, Hudson, Ohio. Available at http://www.lexi.com)

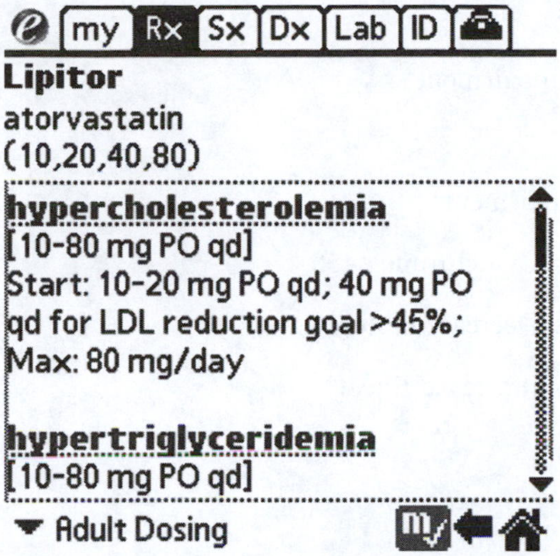

FIGURE 12.9 Epocrates Rx atorvastatin adult dosing. (Epocrates, Inc. 2007. Epocrates Rx Pro, version 8.10. Palm Operating System, San Mateo, California. Available at http://www.epocrates.com)

about the drug, the user clicks on drop-down menus ("Jump" for Lexi-Drugs, the "triangle" in the lower-left corner for Epocrates). Epocrates provides 10 choices:

- Adult Dosing
- Pediatrics Dosing
- Black Box Warnings
- Contraindications/Cautions
- Drug Interactions
- Adverse Reactions
- Safety/Monitoring
- Pharmacology
- Manufacturer/Pricing, and
- a Notes Section (to allow custom user notes to be entered into the monograph).

In contrast, Lexi-Drugs provides 24 choices:

- Adverse Reactions
- Brand Names U.S.*
- Contraindications
- Dosing: Adults

- Dosing: Elderly

- Dosing: Hepatic Impairment*

- Dosing: Pediatric

- Dosing: Renal Impairment*

- Drug Interactions: Cytochrome p450*

- Drug Interactions: Decreased Effect*

- Drug Interactions: Increased Effect*

- Generic Available*

- Lactation*

- Mechanism of Action*

- Medication Safety Issues

- Name

- Pharmacokinetics/Dynamics*

- Pharmacologic Category

- Pregnancy Risk Factor

- Pricing*

- Pronunciation

- Strengths/Dosage Forms*

- Uses, and

- Warnings/Precautions.

The asterisk denotes information that is also available in Epocrates but is grouped within other headings (e.g., renal dosing is grouped with adult dosing). The information in Epocrates Rx is often less detailed than that in Lexi-Drugs.

As mentioned earlier, one of the major differences between the two programs is how information is displayed. Drug interactions for atorvastatin are shown in Figures 12.10 and 12.11. Lexi-Drugs has information categorized according to "increased effect" and "decreased effect" and displays in paragraph style; Epocrates Rx is categorized into "contraindicated," "avoid/use alternative," "monitor/modify treatment," and "caution advised." Additionally in Epocrates Rx, clicking on any named drug interaction brings up a pop-up window with further information or a suggested dosage adjustment, if appropriate. In terms of information, Lexi-Drugs provides more detailed data. Comparing these atorvastatin categories, Lexi-Drugs and Epocrates present the following information:

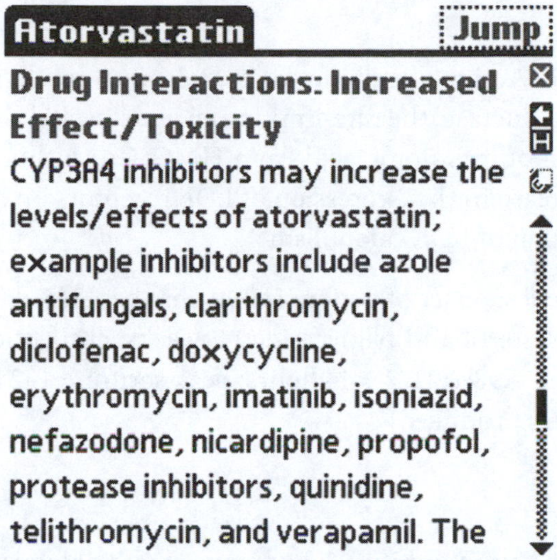

Atorvastatin
Jump
Drug Interactions: Increased Effect/Toxicity
CYP3A4 inhibitors may increase the levels/effects of atorvastatin; example inhibitors include azole antifungals, clarithromycin, diclofenac, doxycycline, erythromycin, imatinib, isoniazid, nefazodone, nicardipine, propofol, protease inhibitors, quinidine, telithromycin, and verapamil. The

FIGURE 12.10 Lexi-Drugs atorvastatin drug interactions. (Lexi-Comp, Inc. 2007. Lexi-Drugs Essentials, file date December 28, 2007. Palm Operating System, Hudson, Ohio. Available at http://www.lexi.com)

my
Rx
Sx
Dx
Lab
ID
Lipitor
atorvastatin
Contraindicated
azole antifungals
Avoid/Use Alternative
amiodarone
clarithromycins
conivaptan
cyclosporine
danazol
Drug Interactions

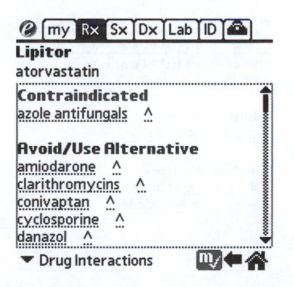

FIGURE 12.11 Epocrates Rx atorvastatin drug interactions. (Epocrates, Inc. 2007. Epocrates Rx Pro, version 8.10. Palm Operating System, San Mateo, California. Available at http://www.epocrates.com)

- Lexi-Drugs

  - Mechanism of Action: inhibitor of 3-hydroxy-3-methylglutaryl coenzyme A (HMG-CoA) reductase, the rate-limiting enzyme in cholesterol synthesis (reduces the production of mevalonic acid from HMG-CoA); this then results in a compensatory increase in the expression of LDL receptors on hepatocyte membranes and a stimulation of LDL catabolism

  - Pharmacokinetics: onset of action: initial changes: 3–5 days; maximal reduction in plasma cholesterol and triglycerides: 2 weeks; absorption = rapid; V_d = 318 L; protein binding ≥ 98%; t1/2 = 14 hours; peak serum = 1–2 hours (also names specific metabolites produced)[3]

- Epocrates Rx

  - Mechanism of Action: inhibits 3-hydroxy-3-methylglutaryl-coenzyme A (HMG-CoA) reductase

  - Pharmacokinetics: t1/2: 14 hours (does not list specific metabolites but does state they are active)[4]

Epocrates Rx also includes two more useful features at no charge: a multiple drug interaction checker (Figure 12.12) and health insurance formularies. The drug interaction checker allows the entry of multiple drugs to check for drug interactions or IV compatibility. Lexi-Comp also offers this feature but at an additional cost. Interactions are again grouped by severity, but rather than listing single drugs, the interacting drugs are listed side by side under the severity. Tapping on the interaction pops up a window with further

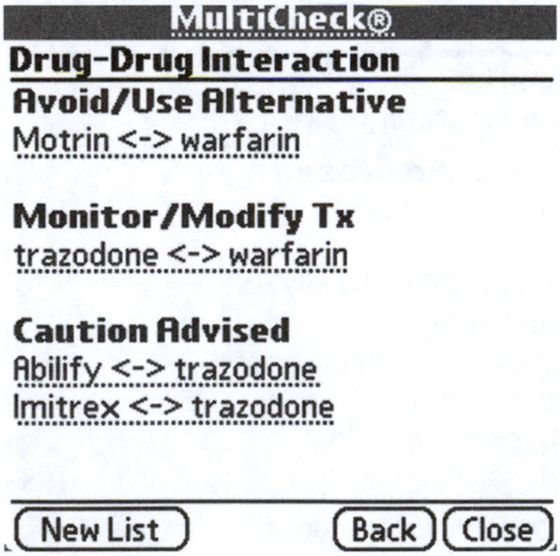

MultiCheck®

Drug–Drug Interaction

Avoid/Use Alternative
Motrin <-> warfarin

Monitor/Modify Tx
trazodone <-> warfarin

Caution Advised
Abilify <-> trazodone
Imitrex <-> trazodone

(New List)   (Back)(Close)

FIGURE 12.12   Epocrates Rx MultiCheck drug interaction analyzer. (Epocrates, Inc. 2007. Epocrates Rx Pro, version 8.10. Palm Operating System, San Mateo, California. Available at http://www.epocrates.com)

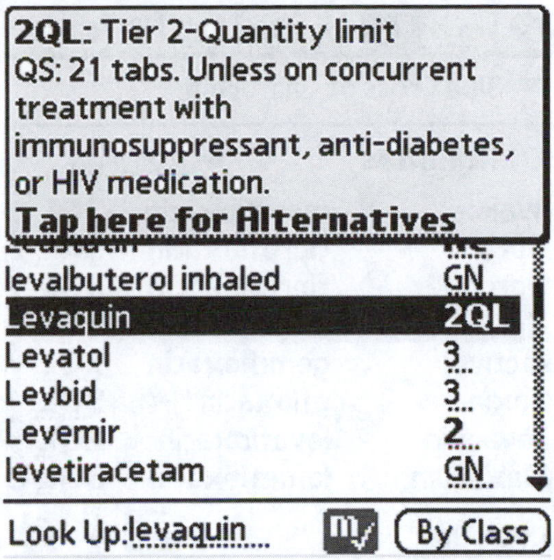

FIGURE 12.13　Epocrates Rx BlueCross of California formulary restrictions for Levaquin. (Epocrates, Inc. 2007. Epocrates Rx Pro, version 8.10. Palm Operating System, San Mateo, California. Available at http://www.epocrates.com)

information. An evaluation of eight PDA drug–drug interaction analyzing programs found the Epocrates analyzer to be the most reliable in detecting interactions.[6]

The health insurance formularies function (not offered by Lexi-Comp) can be particularly useful to pharmacists in community pharmacy settings. Users choose the formularies of interest at the Epocrates Web site; then, the formularies are synchronized to the PDA. Many, but not all, formularies are available. Once turned on, an individual formulary filter causes the Epocrates Rx main screen to display a column on the screen that lists the formulary tier for each drug (see Figure 12.13). Clicking on "2QL" to the right of Levaquin provides a pop-up window describing the formulary restrictions for that drug. Clicking on the "Tap here for Alternatives" of the pop-up window leads to a list of all other drugs in the same class and their formulary status (Figure 12.14).

In summary, both programs are very well designed for pharmacist use. Lexi-Drugs is the more detailed reference. For common information (dosages, drug interactions, adverse reactions), both databases provide adequate information, and it is likely that the user will find it more quickly using Epocrates Rx because of its list-like organization. Considering that it is available at no cost and has a drug interaction analyzer makes it a very useful package. On the other hand, if the user requires more detailed facts—as might be needed in a hospital or educational setting—Lexi-Drugs is a better choice due to its depth of information. Also, both Epocrates Rx and Lexi-Drug could be installed on the same PDA device to take advantage of both.

At least eight other pharmacy PDA drug information programs are available with varying strengths and weaknesses. Individuals and students associated with institutions that subscribe to Thomson Healthcare Series (Micromedex) or Clinical Pharmacology can download the respective PDA versions of these programs free of charge.[7,8]

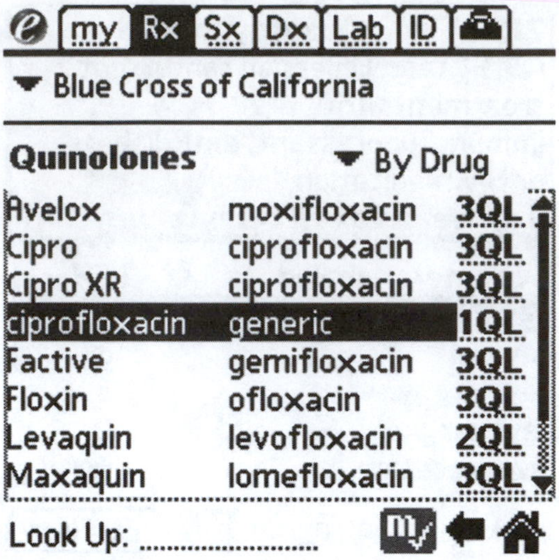

FIGURE 12.14   Epocrates Rx result of clicking on "tap here for alternatives" in Figure 12.13. (Epocrates, Inc. 2007. Epocrates Rx Pro, version 8.10. Palm Operating System, San Mateo, California. Available at http://www.epocrates.com)

12.5 OTHER PDA PROGRAMS USEFUL TO PHARMACISTS

The ability to carry one or more complete drug references in a small, pocket-sized device makes a PDA very useful to pharmacists and other healthcare professionals. Adding other supplemental references to a PDA can enhance its utility. Disease states, laboratory values, medical calculators, dosage form identifiers, complementary and alternative medications, infectious diseases, and medical dictionaries are some examples of other types of applications that are available. These references are offered as stand-alone programs or may be purchased as part of a package, depending on the publisher:

- *Disease state references.* In a disease state program, medical conditions are listed alphabetically and are searchable in the same way as for the drug references. These are offered as upgrades to Epocrates Rx and as stand-alone databases from Lexi-Comp and other publishers. Additionally, there may be an ability to sort conditions by body system (e.g., only displaying conditions that affect the gastrointestinal system). Clicking a condition provides a monograph with information on incidence, causes, differential diagnosis, and treatments including medications. If the disease reference database is associated with a drug reference, clicking an underlined drug name will hyperlink to the drug's monograph for dosage information. These types of references can be useful for reviewing information about a condition when a treatment is to be determined.

- *Laboratory value references.* These databases allow the user to look up reference ranges for a given laboratory test, such as a blood marker or drug concentration, and may even provide suggestions for causes when the value is out of range. This can be a quick and helpful tool to interpret a patient's laboratory test results.

FIGURE 12.15 MedCalc creatinine clearance calculator. (Tschopp, M. 2007. MedCalc, version 5.4. Palm Operating System. Available at http://www.med-ia.ch/medcalc)

- *Medical calculators.* Medical calculators are another useful category of applications for the PDA. They allow quick determinations of values such as creatinine clearance, glomerular filtration rate, ideal body weight, and many others. The user chooses the formula from a predefined list, enters the patient-specific values into the variable fields, and the result is calculated. Often, the program will allow the use of S.I. or local units. Additionally, a reference is typically provided for how the calculation is being made. Although many specialized calculators can be found by searching the Internet, MedCalc is a well-built and free calculator suite.[9] It is designed for Palm, Pocket PC, and the iPhone (Figure 12.15). Over 80 formulas are included. Additionally, it allows the user to create a custom list of most used formulas. It can be found at http://med-ia.ch/medcalc/.

- *Dosage form identifiers.* Surely, every pharmacist has asked a patient what medications he or she takes and the patient responds by presenting a bottle filled with numerous different capsules and tablets. Although not as robust and complete as their online counterparts, identifier programs for the PDA can be useful when at community events or in other settings. Lexi-Comp's Lexi-Drug ID, Epocrates Rx (iPhone and BlackBerry only) and Ident-A-Drug (from the publisher of the *Pharmacist's Letter*) are examples of PDA dosage form identifiers. Both are text-based identifiers that allow the user to input markings and imprints, shape, color, or scoring to identify the unknown drug. Ident-A-Drug and Epocrates even show a picture to help confirm the identity of the drug.

- *Alternative or natural products.* With the abundance of alternative medicines and health supplements used by today's consumers, natural medicine references are great to have on a PDA. Lexi-Comp and Epocrates offer alternative medicine databases, as do others. Epocrates integrates the alternative medicines directly into the list of drugs and also allows the user to include these in drug interaction checks. It gives reported

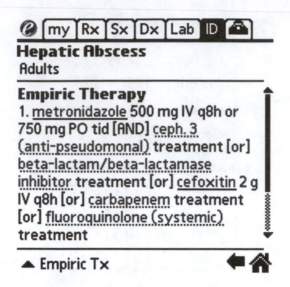

FIGURE 12.16   Epocrates infectious disease. (Epocrates, Inc. 2007. Epocrates Rx Pro, version 8.10. Palm Operating System, San Mateo, California. Available at http://www.epocrates.com)

uses and dosages as well as potential drug interactions and adverse effects. Lexi-Comp's Lexi-Natural Products does not integrate the alternative medicines with the drugs, but is displayed in a paragraph format. The information tends to be more detailed than that in Epocrates and also provides some references. However, perhaps the best alternative medicine program for PDAs is the Natural Medicines Comprehensive Database. Users familiar with the online version will find it to be organized the same way. Efficacy ratings are provided in the categories "likely effective," "possibly effective," "possibly ineffective," and "insufficient evidence." Safety ratings are given in a similar type of tiered classification. It also includes a drug-alternative medication interaction checker and even lists actual brand product names with ingredients.

- *Infectious diseases.* Similar to a disease state reference, these databases allow the user to search for information by infectious agent, syndrome, or body system affected. The program then gives recommended drugs for both empiric and specific treatment (Figure 12.16). Both Lexi-Comp and Epocrates offer versions, but perhaps the most complete and useful is the PDA version of the *Sanford Guide to Antimicrobial Therapy.* The PDA version is essentially an electronic version of the printed guide with all of the same tables and information. Like the printed version, the PDA guide is updated once a year. A study comparing six infectious-disease PDA programs in a hospital setting was conducted by a general medicine teaching service. It found that five out of the six programs provided treatment answers for over 90% of the hospitalized patients with infectious diseases during the study period.[10]

- *Medical dictionaries.* Available from several publishers for both Palm and Pocket PC devices, medical dictionaries are identical to their printed counterparts but have the added advantage of being searchable. Additionally, some dictionaries allow clicking

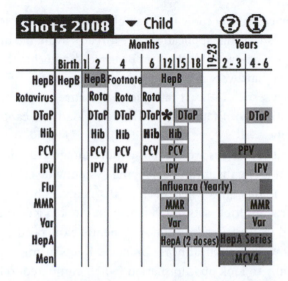

FIGURE 12.17   Shots 2007 child vaccination chart. (Society of Teachers of Family Medicine. 2008. Shots 2008, version 9.0. Palm Operating System. Available at http://www.immunizationed.org/)

on words within the definition to view their meaning. Finding words using a PDA is much quicker than flipping through a printed dictionary. The cost of a PDA-based medical dictionary is comparable to that of a printed version, so it is recommended that PDA owners consider using an electronic version instead.

- *Immunizations.* Another useful, free program worth trying is Shots 2008 (Figure 12.17). It is a full vaccination schedule for children and adults. Based on U.S. recommended childhood and adult immunization schedules, it even gives the recommended catch-up schedule for those missing vaccinations in a series. Clicking on the vaccine name gives further information on the vaccine, such as indications, adverse reactions, con-traindications, and administration. The program has been updated yearly since 2001 and can be downloaded at www.immunizationed.org.[11]

12.6 OTHER USES

PDAs can be used to streamline work flow in other settings. Using a mobile version of Microsoft Excel or even custom-created programs, a PDA can be an excellent tool to collect standardized data that would normally be collected on paper and then tediously entered into a computer. The collected data can be exported to other programs for desktop computer use and analysis. A review on the use of PDAs to document pharmacist interventions discusses some of the benefits and limitations of using PDAs for this purpose.[12] An example of such a use could be in hospital unit inspections for The Joint Commission requirements on medication storage. Using purchased or in-house created software, data can be collected during the inspection and then used to generate a report or even automatic alerts for units that are not meeting requirements.

With the increase in computing power offered by devices such as the iPhone and the ease of developing complex applications that can be quickly and widely distributed, these types of handheld devices will become increasingly incorporated into our lives and work flow. At the time of this writing, applications are already available to allow users to view complex data sets such as a series of MRI scans or three-dimensional models of molecules. Additionally, peripheral hardware is being developed that allows the capture of data from a whole range of devices, such as heart rate, blood pressure, and even bar codes or RFID (radio-frequency identification) tags with ability to upload the captured information to databases automatically.

It is clear that handheld devices are becoming more integrated into the daily practice of healthcare—whether it is the institution deploying systemwide changes or the individual using technology to enhance his or her practice. PDAs are lightweight and portable and offer the ability to carry multiple sets of references in one device. The use of streamlined interfaces, search/find features, and drop-down menus can speed up information usage and enhance work flow in daily healthcare practice. Keep in mind that it can take new users more time initially to look up information (when compared to familiar paper references) due to unfamiliarity with the PDA's interface and software. However, users usually become quite efficient after learning how to use PDAs and their software properly. PDAs are useful reference tools, but they can be equally useful in helping organize our daily lives, for which purpose they were initially designed. Those who use a PDA in both personal and professional lives will get the most out of it.

REFERENCES

1. Palm, Inc. (2002). Palm OS Garnet (version 5.4.0) (Palm operating system). Sunnyvale, CA (available at http://www.palm.com).
2. Microsoft, Inc. (2007). Windows Mobile Today Screen (version 6) (Windows mobile operating system). Redmond, WA (available at http://www.windowsmobile.com).
3. Lexi-Comp, Inc. (2007). Lexi-Drugs essentials (file date December 28, 2007) (Palm operating system). Hudson, OH (available at http://www.lexi.com).
4. Epocrates, Inc. (2007) Epocrates Rx Pro (version 8.10) (Palm operating system). San Mateo, CA (available at http://www.epocrates.com).
5. Epocrates, Inc. (2007). Epocrates privacy policy (available at http://www.epocrates.com/company/privacy.html). Retrieved December 20, 2007.
6. Perkins, N. A., Murphy, J. E., Malone, D. C., and Armstrong, E. P. (2006). Performance of drug–drug interaction software for personal digital assistants. *Annals of Pharmacotherapy* 40(5): 850–855.
7. Thomson Healthcare. (2007). Thomson Clinical Xpert (available at http://www.micromedex.com/products/clinicalxpert). Retrieved December 28, 2007.
8. Gold Standard. (2007). Clinical Pharmacology OnHand (available at http://www.clinicalpharmacologyonhand.com). Retrieved December 28, 2007.
9. Tschopp, M. (2007). MedCalc (version 5.4) (Palm operating system) (available at http://www.med-ia.ch/medcalc).
10. Burdette, S. D., Herchline, T. E., and Richardson, W. S. (2004). Killing bugs at the bedside: A prospective hospital survey of how frequently personal digital assistants provide expert recommendations in the treatment of infectious diseases. *Annals of Clinical Microbiology and Antimicrobials* October 22:3–22.
11. Society of Teachers of Family Medicine. (2008). Shots 2008 (version 9.0) (Palm operating system) (available at http://www.immunizationed.org/).

12. Fox, B. I., Felkey, B. G., Berger, B. A., et al. (2007). Use of personal digital assistants for documentation of pharmacist's interventions: A literature review. *American Journal of Health-System Pharmacy* 64:1516–1525.

APPENDIX 12.1: SUMMARY OF PDA SOFTWARE OF INTEREST TO PHARMACISTS

- Epocrates

  Available: www.epocrates.com

  Devices: Palm, Pocket PC, Windows Smart Phone, iPhone, BlackBerry

  Number of drugs: >3,300

  Updates: daily

  Products/cost:

  > Epocrates Rx: free; drug monographs, formularies, drug interaction checker, calculators

  > Epocrates Rx Pro: $60/year; Epocrates Rx, infectious diseases, alternative meds, IV compatibility

  > Epocrates Essentials: $149/year; Epocrates Rx Pro, disease lists, symptom assessor, lab values

  > Epocrates Essentials Deluxe: $199/year; Epocrates Essentials, ICD-9 codes, medical dictionary

- Lexi-Comp

  Available: www.lexi.com

  Devices: Palm, Pocket PC, iPhone, BlackBerry

  Number of drugs: >7,000

  Updates: multiple times per week

  Free trial: 20 uses of any database

  Products/cost: $75/year to $300/year depending on package and device

  > Lexi-Drugs

  > Pediatric Lexi-Drugs

  > Lexi-Natural Products

  > Lexi-Infectious Diseases

  > Lexi-Lab & Diagnostic Procedures

Lexi-Interact

Lexi-Pharmacogenomics

Lexi-CALC (medical calculators)

Lexi-I.V. Compatibility

Lexi-Drug ID

Harrison's Practice (disease management tool)

Stedman's Medical Dictionary

Bundled packages:

Lexi-Complete: all 21 Lexi databases

Lexi-Select: all 17 drug related databases

Lexi-Clinical: Lexi-Drugs, Interact, Lab & Diagnostic, CALC, Harrison's

- Clinical Pharmacology OnHand

Available: www.clinicalpharmacologyonhand.com

Devices: Palm, Pocket PC

Number of drugs: >6,000

Updates: daily

Free trial: no (free of charge to individuals associated with Clinical Pharmacology subscribing facility)

Products/cost:

Clinical Pharmacology OnHand: $99/year; drug monographs, interaction checker

Clinical Pharmacology OnHand Drug Identifier: $39/year; text-based dosage form identifier

Clinical Pharmacology OnHand IV Alert: $49/year; IV drug compatibility

- Tarascon PDA Pharmacopoeia

Available: http://www.tarascon.com/

Devices: Palm, Pocket PC, BlackBerry

Number of drugs: ~4,500

Updates: daily

Free trial: 30 days free

Products/cost: $29.95/year; PDA version of the popular pocketbook; 47 reference tables, nine medical calculators

- A to Z Drugs

  Available: www.skyscape.com

  Devices: Palm, Pocket PC, SmartPhone, iPhone, BlackBerry

  Number of drugs: 700 generic and 2,800 trade names (source is Facts and Comparisons)

  Updates: quarterly

  Free trial: Web based

  Product/cost: A to Z Drugs: $49/year

- Natural Medicines Comprehensive Database

  Available: www.naturaldatabase.com

  Devices: Palm, Pocket PC

  Updates: daily

  Number of drugs: >1,000 natural and alternative medicines

  Free trial: no

  Product/cost: $59/year; information on drug interactions, uses, and adverse effects; brands, ingredients, and manufacturers listed; ratings on efficacy by indication (possibly effective, possibly ineffective, insufficient evidence to rate); ratings on safety (possibly safe, possibly unsafe, likely unsafe) with references; mechanism of action with references

- MedCalc—medical calculator

  Available: http://www.med-ia.ch/medcalc/desc.html

  Devices: Palm, Pocket PC, iPhone

  Updates: new versions released periodically

  Product/cost: free; >80 formulas sorted by categories; most with bibliographic references and clinical tips

- Sanford Guide to Antimicrobial Therapy

  Available: www.sanfordguide.com

  Devices: Palm, Pocket PC, BlackBerry

  Updates: annually

Free trial: no

Product/cost: $29/year; searchable version of the complete *Sanford Guide to Antimicrobial Therapy*

- Shots 2008

    Available: www.immunizationed.org

    Devices: Palm*, Pocket PC

    Updates: yearly

    Product/cost: free; interactive vaccination schedule for adults and children and catch-up; lists side effects, contraindications, and further information about each vaccine

Note: Information up to date as of spring 2009.

---

* Palm is a trademark of Palm, Inc.

IV

Decision Support

CHAPTER 13

The Practice of Evidence-Based Medicine

Laura J. Nicholson

CONTENTS

13.1 INTRODUCTION

13.1.1 What Is Evidence-Based Medicine?

Evidence-based medicine (EBM) combines the practitioner's clinical expertise and the patient's desires and values with the conscientious and judicious use of the current best evidence from the clinical literature[1] (Figure 13.1). Clinical expertise is one's ability to identify

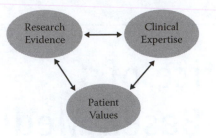

FIGURE 13.1   The three components of evidence-based medicine.

the patient's diagnosis, risks and benefits, and specific clinical circumstance. Patient values refer to the individual patient's concerns, expectations, and preferences that must be integrated into clinical decisions. Current best evidence means high-quality, relevant research, usually from clinical trials, that replaces previously accepted treatment standards with those that are more efficacious and safe.

Although we begin to accumulate medical knowledge during training, unfortunately for us, it is quickly and regularly outdated by newer evidence. Specifically, drugs are studied in additional clinical settings to identify their most beneficial uses, and newer pharmacological agents are introduced frequently, sometimes to supersede prior treatment modalities. EBM acknowledges this problem of eroding medical knowledge, combined with the average practitioner's limited time for obtaining new information, and provides a framework for finding the best available evidence to apply to patient care. The importance of this goal and the resultant EBM movement was first described by Dr. Gordon Guyatt's group at McMaster University in 1992.[2] In this chapter, we will follow their system to demonstrate how to practice EBM in as little as a few minutes per day to find and assimilate new evidence regularly into clinical care.

13.1.2 The Five A's of Evidence-Based Practice

The practice of EBM during patient care is conducted by the following steps, described as the five A's:

1. *Assess* the patient or information need

2. *Ask* a focused clinical question

3. *Acquire* the best available evidence to answer the question

4. *Appraise* the quality of the evidence

5. *Apply* the new information to patient care

Throughout this chapter, we will use clinical cases and follow these five A's to find new evidence to apply to patient care. Although there are many types of clinical questions, such as those about screening, diagnosis, prognosis, economic analysis, and more, this chapter will focus on questions about treatments—more specifically, pharmacological treatments.

13.2 ASSESSING THE PATIENT

Let us use a real clinical case to see how to turn a clinical question into the acquisition of new evidence. This information will then update our personal knowledge base and be useful for future similar patients.

> A 37-year-old female patient is admitted to the hospital with fever, jaundice, and a tender liver after a 1-week alcohol binge. She is diagnosed with alcoholic hepatitis. You are on rounds with the medical team caring for the patient, and they report that the liver specialist has recommended pentoxifylline therapy for this patient. The physicians have not used this treatment before and want to know whether it will really decrease her risk of mortality. They ask you to research this further and bring information to the team.

The first step of the evidence-based practice exercise is to assess the patient. The team has done this together and has made the diagnosis of alcoholic hepatitis. During this assessment, they have identified an information need regarding pentoxifylline use in the setting of alcoholic hepatitis. In order to find an answer, the next step of evidence-based practice is to build a focused clinical question for our search.

13.3 ASKING FOCUSED CLINICAL QUESTIONS: THE PICO FORMAT

An important part of an efficient EBM exercise is to build a focused question out of the clinical need so that the search results are concise and the obtained information does not overwhelm the searcher. In order to do this, the PICO format is used:

- P—a word or phrase for the *patient* or clinical *problem*
- I—the *intervention*
- C—any *comparison* intervention being considered and
- O—*outcome* of concern

The word or words chosen in the PICO format are the ones typed into the search box of the information resource in order to find applicable evidence.

> Sitting down at your computer to begin a search for the requested information, you write out the question: Does pentoxifylline reduce mortality in patients with alcoholic hepatitis? In your PICO format, the terms become: P—alcoholic hepatitis; I—pentoxifylline; C—(none for this case); O—mortality. You type these words into a PubMed search and come up with over a dozen articles. In addition to perusing these publications, you know of a few other information resources that might address this question, and you would like to try your search in those. On the other hand, you may end up with too much to read. Which resource will give you the highest quality evidence? How can you limit your search so that you will not be overwhelmed with excess information?

13.4 ACQUIRING EVIDENCE: DIFFERENT RESOURCES
FOR DIFFERENT INFORMATION NEEDS

Each of us has a few information resources we regularly use. We use some of them for very general questions, such as reading about a rare disease for the first time in a textbook or online resource such as UpToDate. We consult others for more specific questions, such as aspirin for stroke prevention in a clinical trial from PubMed or in a consensus statement from a national neurology association. In this section, we will discuss the general attributes, strengths, and weaknesses of a few evidence resources, using the preceding clinical case. The focus will be on online databases because they are the most efficiently accessed and frequently updated. The reader is strongly encouraged to try the online searches while reading this chapter, but please keep in mind that results may differ due to new evidence published in the interim.

13.4.1 PubMed

A service of the U.S. National Library of Medicine and free through the Internet, PubMed contains millions of biomedical citations, from MEDLINE and other life science databases, immediately upon their release. Strengths include that it is the most current and comprehensive evidence resource, it has several tools to improve the accuracy of a search, and it is free without subscription. Chapter 5 describes PubMed searching in detail.

> For your alcoholic hepatitis patient, you type into PubMed: "alcoholic hepatitis and pentoxifylline and mortality." You obtain over a dozen articles—too many to read in their entirety and share with the team. Using the "Limits" function, you choose to limit your search to "Randomized Controlled Trial" because this is the best type of therapy trial. Now you have only one citation to read, which appears to match your question perfectly.[3] You decide to review this article and present the results tomorrow.

13.4.2 *ACP Journal Club*

A product of the American College of Physicians (an internal medicine society), *ACP Journal Club* is a journal that reviews over 100 other internal medicine journals in order to publish short summaries of those articles considered by EBM experts to be of the highest quality and greatest importance to the field of internal medicine. Strengths include the ease of reading a short summary rather than an entire article and the expert review provided. Weaknesses include the lag time required for identification and summarization of the important clinical trials and the subscription fee imposed. (Note that some institutions pay the subscription fee, making online access free through their networks.)

> You notice that the article you found was published in 2000 and wonder whether it has been reviewed by *ACP Journal Club* in the interim. You type the same search, "alcoholic hepatitis and pentoxifylline and mortality," into *ACP Journal Club*'s search box and get one result.[4] This is, in fact, an expert review of your article, giving the synopsis and pertinent results! Now you can review the entire clinical trial in just a few minutes, and you can rest assured no newer randomized controlled trial has taken place since you looked in PubMed first.

13.4.3 UpToDate

An online educational product begun in 1989, UpToDate contains thousands of chapters on basic pathophysiology and current treatment recommendations for diseases of interest to internists, family practitioners, gynecologists, and pediatricians. Strengths include its comprehensive and easy-to-read listings with extensive bibliographies. Weaknesses include the high subscription price and the lag time required for its authors to search, summarize, and publish information. (Note that some institutions pay the subscription fee, making online access free through their network.)

> Because you have not seen a case of alcoholic hepatitis before, you decide to review the disease in UpToDate. You first search just the words "alcoholic hepatitis" and come up with several chapters describing the pathophysiology, diagnosis, and treatment of the illness in addition to other aspects of alcoholic liver disease. Briefly scanning, you learn about the seriousness of this problem and its high mortality. Next you add the term "pentoxifylline" and find a chapter containing a statement that one trial found this drug efficacious. You have learned a lot more about alcoholic hepatitis, but not whether pentoxifylline is definitely recommended.

13.4.4 National Guideline Clearinghouse

A comprehensive database of evidence-based clinical practice guidelines, National Guideline Clearinghouse is maintained by the federally funded Agency for Healthcare Research and Quality in order to disseminate clinical practice recommendations to healthcare providers. Strengths include its concise and usually evidence-rated recommendations, as well as that it is free through the Internet at www.guidelines.gov. Weaknesses include the wide variety of guideline authors and the retention of some outdated guidelines.

> It occurs to you that perhaps a society of liver specialists has published a guideline on the care of patients with alcoholic hepatitis. You enter your PICO terms in the search box of National Guideline Clearinghouse, but get no results. Because this search may be too specific, you enter only "alcoholic hepatitis," which produces 11 related guidelines. Reviewing the titles, you see that none specifically addresses your patient's illness, so you abandon this resource for this particular occasion.

13.4.5 Cochrane Reviews

Produced by a nonprofit EBM collaboration, Cochrane Reviews are very high-quality, evidence-based reviews addressing well-researched clinical questions. Most often, multiple randomized controlled trials addressing the same clinical question are combined in a meta-analysis (discussed in a later section of this chapter) to produce these reviews. Strengths include the expert analysis used to produce a single best summary of multiple trials; weaknesses include the time lag required to produce such a high-quality summary, during which new evidence may be published. In this author's opinion, the Cochrane Web site is very complex to search, and the most efficient way to find a Cochrane Review is through a PubMed limit, as illustrated in the following clinical context:

> To check for an expert Cochrane Review on this topic, you type "alcoholic hepatitis" into the PubMed search box and then select the "Limits" function. Under "Limits," you choose "Search by Journal" and type in "Cochrane." Clicking on the choice revealed, which is "Cochrane database of systematic reviews," you then click "Go" to repeat your search and get four different Cochrane Reviews. (Note that a single Cochrane Review may be listed more than once if it has been updated.) Two reviews address other agents for alcoholic liver disease, anabolic-androgenic steroids and propylthiouracil,[5,6] but neither is about pentoxifylline. To see whether steroids or

propylthiouracil helped, you quickly review those two abstracts. Neither had a significant effect on survival according to the Cochrane experts—information you might mention when you present your pentoxifylline data to the team.

13.4.6 Comparison of Resources

This section on acquiring evidence demonstrates how several resources can be searched for a single clinical question to answer different aspects of an information need. PubMed reveals the most recent data and finds the clinical trials that match our question, if they have been done. *ACP Journal Club* summarizes the publication, if its quality and pertinence are high, and UpToDate, National Guideline Clearinghouse, and Cochrane Reviews provide summaries that place the trial in a broader clinical context. The searcher's needs dictate which resource is used; for example, a liver specialist might want only the most recent publications from PubMed, while a pharmacy student might need to review pathophysiology and other treatment options in a resource like UpToDate. With database familiarity, information from all of them can be combined within a few minutes of searching.

Notice that among the resources discussed before, all but PubMed are prescreened, tertiary databases; in other words, the evidence has been synthesized and summarized by others (see Section 5.2 in Chapter 5). This requires a considerable amount of time during which new evidence may become available. For this reason, searches in a prescreened database should typically be followed by a search in PubMed to obtain any more recent publications. Additionally, the reader is probably aware of several resources that have not been discussed previously. The choices presented here are based on their wide availability and complementary attributes. The practice of EBM optimally includes an effort to be aware of new evidence resources as they become available.

13.5 APPRAISAL: MAKING SURE THE EVIDENCE IS GOOD

It can be tempting to find a published article, read the abstract, and immediately apply it to patient care. The risk with this approach is that a substandard publication or clinical trial might be used to make important treatment decisions because, unfortunately, articles with important methodological errors are published frequently. How then do we make sure that we are using the best available evidence for patient care? The simplest way is to use evidence from a trusted resource that has been prescreened by EBM experts. This is exactly the reason for the existence of *ACP Journal Club*, Cochrane Reviews, and other preappraised resources. The authors of these products understand that many clinicians do not have the time or skills to filter the best evidence from the enormous quantity of available literature. They do it for us so that we can efficiently and confidently find high-quality evidence for our clinical decisions.

What about articles that have just been published and have not yet appeared in prescreened resources or a trial that, for some reason, was not included in a preappraised summary? How can we determine the quality of evidence in these publications? The answer is provided in the *Users' Guides to the Medical Literature*,[7] published to help us, the users of the medical literature, evaluate the quality of the evidence contained therein. The *Users'*

Guides to the Medical Literature is available at www.cche.net/usersguides/main.asp at no charge or in an interactive format at www.usersguides.org by paid subscription. These sites include a chapter (a "guide") for each article type. We will focus on articles about therapy, the type that includes drug trials.

13.5.1 Therapy Article: The Randomized Controlled Trial

Trials that compare several drugs or a single drug to placebo are known as randomized controlled trials. This title refers to two important methodological aspects of a good drug trial: randomization and a control group. Other important conditions make a drug trial valid, and we will review these using our previous clinical example:

> Imagine that the year is 2000 and the article you obtained through your PubMed search of "alcoholic hepatitis and pentoxifylline and mortality," limited to randomized controlled trials, was just published a few weeks ago.[3] You would like to take this information to the medical team, but you must first make sure it is good evidence. Opening the free online *Users' Guides to the Medical Literature* (at www.cche.net/usersguides/main.asp), you choose the chapter entitled "Therapy or Prevention." Skipping the sections on the clinical scenario and search because you already have your own, you proceed to the three-step instructions for appraisal. With your article in hand, you open the section: "I. Are the results of this article valid?"

Table 13.1 lists the methodology questions that must be asked of a therapy trial as the first step in the appraisal process. These questions check that a drug trial has been done with the least possible bias and thereby determine the trial's level of validity. The meaning and purpose of each question are explained in the corresponding section of the online *Users' Guides to the Medical Literature*. Once the reader is satisfied with the trial's answers to these questions and believes that the study's authors minimized bias wherever possible, only then should the results be considered.

Interpreting results is the second step in the appraisal process for an article about therapy. To interpret a study's results, the online *Users' Guides to the Medical Literature* contains the corresponding section, "II. What were the results?" This section provides comprehensive calculations and explanations for how to use a trial's results to predict the magnitude of clinical effect of the drug (or other therapy) being studied. Often an article presents some, but not all, of these calculations and the reader must compute the rest for a full interpretation. The equations are summarized in Table 13.2 and used in the following sample calculations for the patient previously described:

TABLE 13.1 ■ Articles about Therapy: Validity Questions

- Were patients randomized?
- Were patients accounted for at the trial's conclusion?
- Was follow-up complete?
- Were patients analyzed in the groups to which they were randomized?
- Were patients, practitioners, and study personnel "blinded" to treatment?
- Were groups similar at the start of the trial?
- Aside from the intervention, were the groups treated equally?

TABLE 13.2   Articles about Therapy: Result Calculations

- Risk without drug: control subjects with outcome/total in control group = control event rate (CER)
- Risk with drug: experimental subjects with outcome/total in experimental group = experimental event rate (EER)
- Relative risk (RR) = EER/CER
- Relative risk reduction (RRR) = 100% − RR, or = (CER − EER)/CER
- Absolute risk reduction (ARR) = CER − EER, or EER − CER if EER is larger
- Number needed to treat (NNT) = 1/ARR

Satisfied with the answers to the validity questions for the article by Akriviadis,[3] you proceed to the result calculations. For the outcome of mortality, the difference between the pentoxifylline and placebo groups was significant, represented by a *p*-value* of 0.037. The risk of death in the placebo group (CER) was 24/52 or 46%. The risk of death in the pentoxifylline group (EER) was 12/49 or 24%. Relative risk (RR) was therefore 24%/46% or 52%, meaning a patient taking pentoxifylline had only 52% the mortality risk of a patient taking placebo. Another way to express this comparison is using relative risk reduction (RRR)—here, 100% − 52% = 48%, meaning a patient taking pentoxifylline had 48% *less* risk of death than someone taking placebo. You are excited to share with your team that your patient's risk of death will drop approximately in half if pentoxifylline is given, according to this study.

To get closer to the clinical bottom line and determine how much you can change mortality for this type of patient, you decide to compute absolute risk reduction (ARR) and number needed to treat (NNT). ARR is 46% − 24% = 22% for this trial, meaning that 22% of a population of alcoholic hepatitis patients will avoid mortality if they receive pentoxifylline. NNT is then 1/0.22 = 4.5. This means that for every five alcoholic hepatitis patients who receive pentoxifylline, one death will be prevented. You are anxious to share these results and now believe that pentoxifylline could help prevent death in your patient, who has a very serious illness as evidenced by the CER of 46%. You draw Figure 13.2 to help explain your findings.

Note that populations with the same relative risk reduction may have very different numbers needed to treat. In a fictitious example, if aspirin reduces heart attack by 50% in all age groups, but 50-year-olds have a baseline risk of 2% for heart attack and 80-year-olds have a baseline risk of 20%, then we have to treat one hundred 50-year-olds but only ten 80-year-olds to prevent one heart attack in each age group. (In 50-year-olds, CER is 2%, EER is 1%, and ARR is 1%, making NNT 100. In 80-year-olds, CER is 20%, EER is 10%, and ARR is 10%, making NNT 10. In both groups, RRR is 50%.)

Contrary to the preceding examples, some studies measure good outcomes, such as successful smoking cessation. In these cases, results are calculated in the same way, but nomenclature changes. Instead of relative risk reduction, the comparison becomes relative benefit increase. For example, with a new imaginary product called CigFree, 40% of prior

---

* The p-value is the probability that a difference will be observed between two groups when they are in fact the same. In our example, a *p*-value of 0.037 means that if there is in truth no mortality effect for alcoholic hepatitis patients treated with pentoxifylline, the trial's apparent difference of 46% and 24% would occur by chance 3.7% of the time. By convention, we accept a *p*-value ≤ 5%, meaning a ≤5% chance that we are assuming a difference when there is none. Statisticians choose among several different equations to compute *p*-values, depending on the population tested and the type of comparison made. These are beyond the scope of this chapter, but it is important to know that the greater the number of subjects is in a trial, the smaller is the chance that we are assuming an effect when there is none (i.e., higher subject numbers produce smaller *p*-values for a given measured difference).

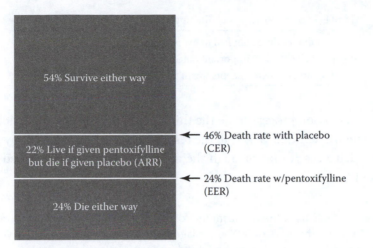

FIGURE 13.2   Illustration of the concepts of control event rate (CER), experimental event rate (EER), and absolute risk reduction (ARR). Number needed to treat (NNT) is the number of times that a drug must be given to the entire population in order to catch one patient in the ARR group, the group who will experience a changed outcome from receiving the drug.

smokers were still abstinent at 1 year, compared with 10% 1-year abstinence among similar subjects taking placebo. Calculating EER/CER gives a relative benefit (RB) of 40%/10%, or 400%, and a relative benefit increase (RBI) of 400% – 100%, or 300%. Absolute benefit increase (ABI) is 40% – 10%, or 30%, making the number needed to benefit (NNB) equal to 1/0.3, or 3.3. (Of course, NNB is 4 because it is impossible to treat 3.3 people.) In addition, some interventions increase bad outcomes rather than reducing them. In this situation, the number needed to harm is the final result calculation, demonstrated in Section 13.7.

It is important to understand that the preceding calculations apply only to studies measuring yes/no outcomes, such as death, pregnancy, or hospitalization. They cannot be applied to continuous outcomes such as bone density, weight, or days in the hospital. Sometimes authors will make continuous outcomes into yes/no outcomes by assigning a threshold value. For example, to measure worsening renal function, a trial may record an outcome of reaching twice one's baseline creatinine, a yes/no event, rather than comparing mean creatinine values between groups.

Another strategy is to use a clinically relevant yes/no outcome that might be predicted by a continuous measure. Such an example would be an osteoporosis intervention trial that compares fracture events rather than bone density in each study arm. By using an absolute outcome that either occurs or does not, authors allow us to interpret and understand the magnitude of their results using RR, RRR, ARR, and NNT, as described here and in the *Users' Guides to the Medical Literature.*

13.6 APPLYING EVIDENCE TO PATIENT CARE

Once a clinical trial has been appraised and its results accepted, how it applies to a patient must be considered. An outstanding article on an efficacious new drug may neverthe-less not apply to a particular patient. The online *Users' Guides to the Medical Literature*

TABLE 13.3   Articles about Therapy: Applying Results to Patient Care

- Can the results be applied to my particular patient?
- Were all clinically important outcomes considered?
- Are the treatment benefits worth the potential harms and costs?

considers the application process to be the third step in appraising an article, and its corresponding section is entitled "III. Will the results help me in caring for my patients?" The questions that drive the reader through the application step are explained there and are listed in Table 13.3.

> You take your findings back to your team the following day, presenting the impressive RRR for mortality of 48% and the NNT of 5. You explain to them that this means that if you place your patient on pentoxifylline, her risk of death should fall by 48%. In addition, for every five patients like her who are given pentoxifylline, one will avoid death as a result. The team is very impressed. The lead physician wants to make sure that the study you found should really be applied to this patient. Together, you check the inclusion and exclusion criteria and find that she would have been enrolled in this trial had she been available and consented. Next you consider whether the study's outcome is clinically important. Certainly death is clinically relevant! In addition, the study also found that pentoxifylline produced a reduction in the clinically important outcome of hepatorenal syndrome, a major cause of kidney failure in alcoholic hepatitis patients. Finally, your team discusses whether pentoxifylline has the potential to hurt the patient. In the article, some untoward effects, such as gastrointestinal upset and headache, were recorded. However, they occurred in a small number of patients and were not statistically different between the treatment and control groups. Ultimately, you decide that this article applies very well to your patient, and pentoxifylline is ordered for her.

13.7 CAUSING NEGATIVE OUTCOMES: NUMBER NEEDED TO HARM

The preceding case demonstrates the five important steps of evidence-based practice: *assess, ask, acquire, appraise,* and *apply.* It also illustrates an example of the best type of therapy evidence: the randomized, controlled trial. Before we move on to other types of evidence, an additional application issue to discuss is the number needed to harm (NNH). Occasionally, a randomized, controlled trial measures a harmful outcome in addition to a beneficial one, and the harmful outcome occurs more frequently in the therapy arm than in the control arm.

In this entirely fictitious example, assume all outcomes are significantly different ($p < 0.05$) between groups. A new clot-prevention drug is associated with 5% clot recurrence in the drug group compared to 25% in the placebo group. ARR is therefore 20%, and NNT to prevent one clot is 1/0.2, or 5. However, bleeding was observed in 30% of the drug group, but in only 10% of the control group. For the outcome of bleeding, absolute risk increase (ARI) is 20%, making NNH to cause one episode of bleeding also 1/0.2, or 5. This means that for every five patients who receive this drug, one clot will be prevented and one episode of bleeding will be caused. Weighing NNT against NNH is an important part of the *apply* step when harmful outcomes are found to be relevant in randomized, controlled trials.

13.8 TAKING EVIDENCE TO THE NEXT LEVEL: INTEGRATIVE PUBLICATIONS

13.8.1 Overview, Systematic Review, or Meta-Analysis

When the same therapy question has been studied many times over the years or in different locations by different authors, the resultant trials may be combined to produce an "overview" of the topic. Combining subjects from numerous trials should give more power to and thereby more confidence in the overall results. When authors of these overviews use bias prevention methods for selection of the trials included, the publication is called a "systematic review," and the mathematical method used to combine the results of the separate trials into one summary estimate is referred to as a "meta-analysis." These three terms are often used interchangeably, and we will use "meta-analysis" here.

In PubMed, if several randomized, controlled trials are found by the searching methods described earlier, it is then wise to look for a meta-analysis. The authors of the meta-analysis have read all of these studies in great detail and combined them into a summary result, so the searcher does not have to attempt to do so on his or her own. Reading the meta-analysis is much more efficient than trying to read all of the individual trials, and it is more accurate than simply choosing what might appear to be the largest or the best among many trials. By analyzing and combining data for the reader, meta-analyses become another source of preappraised information in the effort to practice EBM. As discussed previously, because new data may appear during the time it takes to publish preappraised summaries, the searcher must also look for any trial too new to have been included in the meta-analysis.

To find a meta-analysis, limit a PubMed search to "Meta-Analysis" under the "Type of Article" limit, or, as discussed before, a Cochrane Review may be sought. Cochrane Reviews are among the very highest quality meta-analyses available. The Cochrane Web site itself may be searched (if the reader has access) or a Cochrane Review may be obtained through a PubMed search, using the limit "Search by Journal" and typing in "Cochrane." From either resource, if a meta-analysis is obtained, the searcher now has a preappraised summary of multiple trials addressing the therapy in question.

Though the meta-analysis contains studies that have been individually appraised by its authors, the meta-analysis itself should be appraised by the searcher. Bias can occur when writing a meta-analysis, such as combining studies that are not sufficiently similar or including an individual trial of substandard quality (i.e., not randomized or blinded). The details of how to appraise and apply a meta-analysis are beyond the scope of this chapter, but they can be found in the online *Users' Guides to the Medical Literature* (at www.cche.net/usersguides/main.asp), choosing the chapter entitled "Overview."

13.8.2 Clinical Practice Guideline: Adding Recommendations to Evidence

Clinical practice guidelines are defined as "systematically developed statements to assist practitioner and patient decisions about appropriate health care for specific clinical circumstances."[8] A guideline is typically written as a consensus statement by a group of experts in the associated field and may be intended for local use, such as a single hospital, or international use, such as all practitioners taking care of a given disease anywhere in the world. Examples include a practice guideline to prevent venous thrombosis in inpatients, written by a local hospital's appointed committee, or one to prevent strokes in all Americans, written by the American Academy of Neurology. The purpose of clinical practice guidelines is to summarize the best evidence for a clinical condition in order to make care more efficient and consistent among practitioners. The desire is to improve outcomes and lower cost. These guidelines can be incorporated into a clinical decision support system (see Chapter 15).

A practice guideline considers the resources and desires of its intended audience and, in this way, combines evidence with practitioner and patient values to create recommendations for the reader. An extensive review of the literature has been done, and often a meta-analysis is used as part of the supporting evidence. Thus, a clinical practice guideline is yet another type of preappraised evidence to increase our ability to practice EBM.

To find clinical practice guidelines, search PubMed, limiting to "Practice Guideline" under "Type of Article," or search National Guideline Clearinghouse (at www.guidelines. gov), as discussed before. In National Guideline Clearinghouse, guidelines appear in short, outline form and are more efficient to read than the full-length publications to which they are linked. As with all prescreened evidence, the search for a clinical practice guideline should be accompanied by a search for any newer data published in the interim.

To appraise and apply a clinical practice guideline, once again the *Users' Guides to the Medical Literature* is our resource. The online version (at www.cche.net/usersguides/main. asp) contains the relevant chapter entitled "Clinical Practice Guideline." As with meta-analyses, appraisal of practice guidelines is beyond the scope of this chapter, but we will point out a few important characteristics to be used during a search:

Consider the source of the guideline. One written by a society of nurses treating patients in rural Africa may not apply to physicians working in a large European city.

Check the date. An excellent guideline published in 1990 almost certainly will have been made obsolete by newer studies.

Newer guidelines contain a grading system for the quality of the contained evidence and the resultant strength of the recommendations. Look for these grades and for an associated key explaining the guideline's grading system.

For additional guideline appraisal, consult the appropriate *Users' Guides to the Medical Literature* chapter.

13.9 SUMMARY

Evidence-based medicine combines clinical expertise and the patient's needs with the very best available evidence—the three critical components of clinical decision making. Using the five A's framework to practice EBM on a daily basis prevents the use of obsolete therapies and provides a method for ongoing practitioner education. Beginning with the patient allows the clinical situation to drive the process. This prevents the pitfall of beginning with an article, perhaps found by browsing, which is later used for a patient to whom it may not apply. A good PICO search as the second step means that all evidence relevant to the focused question can be reviewed. Skipping the PICO search—again, by starting with an article rather than a clinical scenario—may result in missing a newer trial that supersedes the article in hand.

The *acquire* step, the third step in the framework, is made easier by knowing a few resources well and using them in a complementary fashion. Prescreened resources are often easier to read and use, as long as the issue of lag time is considered. Among the examples given earlier, *ACP Journal Club* may expertly appraise one trial, while PubMed will list any newer trials done in the interim. Combining a search of National Guideline Clearinghouse and of Cochrane Reviews produces any expert guideline written on a topic and the meta-analysis on which the guideline was based. The critical consideration when evidence from any prescreened resource is acquired is to remember to look for any newer evidence published in the interim by searching PubMed as a final check.

The *appraisal* and *apply* steps may be skipped if a trusted preappraised resource is used. However, if the reader wants to review the trial and its quality in great detail (perhaps for a journal club or presentation to colleagues), the *Users' Guides to the Medical Literature* contains chapters detailing appraisal and application of therapy trials, meta-analyses, clinical practice guidelines, and many other article types not discussed in this chapter. It helps the practitioner to check the quality of the evidence and find any potential sources of bias before using literature for patient care. Sometimes, in this way we discover that a standard therapy is based on relatively weak evidence and that a stronger trial should be designed.

The randomized, controlled trial is the best study design for analyzing efficacy of a drug, and it is the type of study pharmacists and physicians will read most often. It is important to recognize the methodological characteristics that make a randomized trial valid and to understand how to express its results. The authors of the study may present only outcome rates and the relative risk reduction, leaving the reader to compute the absolute risk reduction and number needed to treat. Knowing how to compute these and doing so regularly are critical to understanding how to apply a trial to patient care because two populations with the same relative risk reduction may have very different numbers needed to treat (as illustrated previously).

This chapter has presented a method for regularly and quickly proceeding from a patient's therapeutic need to the application of new data, with continuing education for the practitioner along the way. Although randomized, controlled trials are the mainstay of clinical drug information, we should begin and end with patients in order to use these trials properly. When several similar trials have been conducted, experts may combine them in a meta-analysis. When much evidence is available for a given clinical syndrome, experts may produce a clinical practice guideline. Evidence-based clinical practice guidelines improve patient outcomes,[9] which is the ultimate goal of EBM in making decisions about patient care. Becoming familiar with a few well-chosen information resources like those detailed here makes it simpler to use all of these types of evidence in the real-time practice of EBM.

REFERENCES

1. Straus, S. E. *Evidence-based medicine: How to practice and teach EBM,* 3rd ed. New York: Elsevier/Churchill Livingstone, 2005.
2. Evidence-Based Medicine Working Group. Evidence-based medicine. A new approach to teaching the practice of medicine. *JAMA* 1992. 268:2420–2425.
3. Akriviadis, E., Botla, R., Briggs, W., Han, S., Reynolds, T., and Shakil, O. Pentoxifylline improves short-term survival in severe acute alcoholic hepatitis: A double-blind, placebo-controlled trial. *Gastroenterology* 2000. 119:1637–1648.
4. Koretz, R. L. Pentoxifylline improved short-term survival in severe acute alcoholic hepatitis. *ACP Journal Club* 2001. 135:4.
5. Rambaldi, A., and Gluud, C. Propylthiouracil for alcoholic liver disease. Cochrane Database System Review 2005. 4:CD002800.
6. Rambaldi, A., and Gluud, C. Anabolic-androgenic steroids for alcoholic liver disease. Cochrane Database System Review 2006. 4:CD003045.
7. Guyatt, G. *Users' guides to the medical literature: Essentials of evidence-based clinical practice,* 2nd ed. New York: McGraw–Hill Medical, 2008.
8. Field, M. J., and Lohr, K. N. Institute of Medicine. Committee to Advise the Public Health Service on Clinical Practice Guidelines, U.S. Department of Health and Human Services. *Clinical practice guidelines: Directions for a new program.* Washington, D.C.: National Academy Press, 1990.
9. Weingarten, S. Translating practice guidelines into patient care: Guidelines at the bedside. *Chest* 2000. 118:4S–7S.

CHAPTER 14

Clinical Pharmacokinetics Computer Programs

Philip O. Anderson

CONTENTS

14.1 INTRODUCTION

Clinical pharmacokinetics is used to improve drug therapy outcomes by optimizing the drug dosage regimen. Most hospital pharmacies provide some pharmacokinetic consultation services and studies have documented that improved patient outcomes and cost savings result from this activity.[1–7] Although many computerized pharmacokinetic analysis methods have been described, Bayesian methods have the longest proven track records for individualizing dosage regimens for patients at the lowest cost in the clinical setting. This chapter describes the theoretical basis for Bayesian pharmacokinetics and provides criteria to use in judging pharmacokinetics programs.

Drugs that have small margins between therapeutic and toxic serum concentrations (narrow therapeutic indices) or drugs with no readily observable clinical end points are the most in need of dosage regimens tailored to the individual patient. Examples of drugs with narrow therapeutic indices are aminoglycoside antibiotics, warfarin, and digoxin. Additionally, drugs such as lithium and some anticonvulsants cannot be rapidly or safely monitored clinically because they could subject the patient to unnecessary toxicity or treatment failure.

14.1.1 Population Pharmacokinetics

Numerous patient factors can affect a drug's pharmacokinetics. These include the patient's age, sex, height, weight, organ function (especially kidney and liver function), genetic polymorphisms in drug-metabolizing enzymes, concurrent disease states, some treatment modalities (e.g., dialysis, mechanical ventilation), and concurrent medications that interact with the drug in question. Often, not all of these factors are taken into account by prescribers because of the complexity of recalling and applying all of the adjustments to normal population values to tailor initial pharmacokinetic values to the specific patient. However, even when all of these factors are taken into account, some variability still exists in pharmacokinetic parameters because of factors that cannot be accounted for.

Population variability of pharmacokinetic parameters can be distributed in the classic normal distribution (the bell-shaped curve), log-normal distribution, or occasionally in other patterns. What most pharmacists learn and what most pharmacokinetic formulas calculate are the mean values of these distributions. Although these average values are useful for the calculation of initial drug dosage, further individualization of drug dosage can be accomplished by using information from serum drug concentrations obtained from the patient after the drug has been administered.

14.2 INDIVIDUALIZING PHARMACOKINETICS

After obtaining serum drug concentration values in a specific patient, proper analysis of these values can be used to obtain an estimate of the drug's pharmacokinetic parameters in the patient. The clearance (*CL;* volume of plasma cleared of drug per unit of time) and apparent volume of distribution (*Vd;* apparent volume into which the drug distributes) are most often calculated for drugs given by intravenous injection. The bioavailability (*F;* fraction of the dose reaching the systemic circulation) is also needed for accurate dosage regimen calculation of oral medications. The number of properly timed serum concentration values needed to calculate pharmacokinetic parameters accurately for an individual is at least one more than the number of parameters being calculated. Therefore, if one is attempting to calculate the *CL* and *Vd* of an intravenously administered drug, at least three data points are needed.

After obtaining serum concentration values, one could simply graph the points on semilogarithmic graph paper and calculate the *CL* and *Vd* of the drug, but this method is not clinically practical for routine use. Handheld programmable calculators and computer spreadsheet programs (e.g., Excel) generally use the method of linear regression to calculate the slope and intercept. This is shown in Figure 14.1, where time is in hours and serum level is in milligrams per liter.

Values from the curve fitting would be an intercept value of *Vd* of 11.31 L and an elimination rate constant (*kd*) of 0.3466/hour. The clearance would be $kd \times Vd$, or 3.92 L/hour. The linear regression method has some noteworthy mathematical limitations. It does not handle outlier values far away from the curve well, potentially resulting in specious results when an outlier is present. Many different factors, such as improper recording of times or obtaining blood from the vein where the drug is infusing, can result in false values. Linear

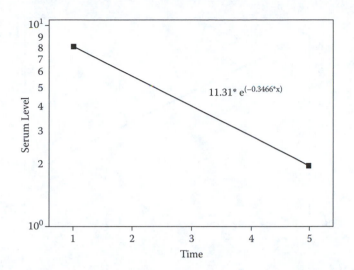

FIGURE 14.1   Linear least-squares fit.

regression also weights low values more heavily (i.e., they affect the results more) than high values, as shown in Figures 14.2 and 14.3.

Note that in these figures, a change in a serum concentration value by 0.2 mg/L near a serum concentration of 2 mg/L causes a greater change in the slope and intercept than a 0.2 mg/L difference near a value of 8 mg/L. The linear regression method has additional pharmacokinetic limitations. First, it disregards all that is known about the specific patient's drug pharmacokinetics based on population (demographic) estimations, disease states, organ function, and drug interactions. Second, serum concentration values must be from the same dose and dosage interval (or assumed to be steady-state values). Values drawn during other dosage intervals may not be reliable because, in clinical practice, dosage intervals are often not equal throughout the day and are therefore not truly at steady state.

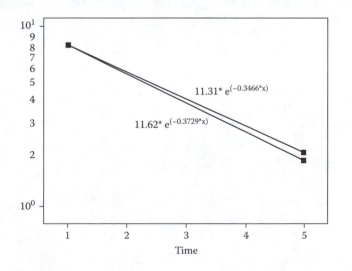

FIGURE 14.2   Low-value difference.

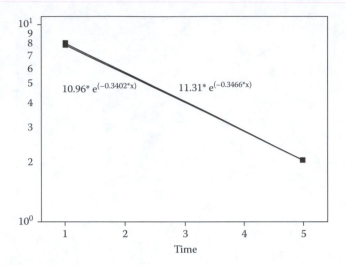

FIGURE 14.3   High-value difference.

14.3 NONLINEAR LEAST-SQUARES FITTING

More sophisticated methods of curve fitting are available using the personal computer. One method is nonlinear regression. This method uses iteration to solve for the pharmacokinetic variables of the specific patient. It is referred to as nonlinear because the curve generated by the process is not a straight line; rather, it is a curved or nonlinear plot because the actual serum concentrations are used rather than logarithms of the values. Figure 14.4 shows an idealized nonlinear curve fitting of six serum concentrations.

In nonlinear curve fitting, an appropriate pharmacokinetic model is first defined for the specific drug—for example, a one-compartment linear model with intermittent intravenous infusions as the method of drug input. Then, a "best guess" estimate of the specific patient's pharmacokinetic parameters (e.g., Vd, CL) is made. These values are often referred to as *a priori* values or "priors" and usually consist of the population values derived from the

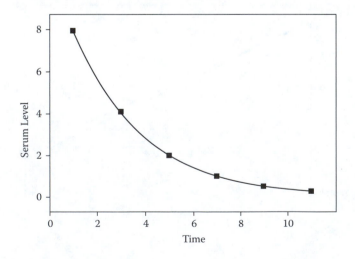

FIGURE 14.4   Idealized nonlinear fit.

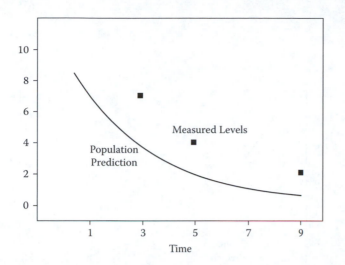

FIGURE 14.5   Starting parameters least-squares fit.

patient's demographics (e.g., age, height, weight), excretory organ function (e.g., serum creatinine), and any other factors known to affect the drug's pharmacokinetic parameters.

These values are entered into the mathematical model to generate expected concentrations at the times that serum drug concentrations were measured. There is usually a difference between the predicted and actual serum concentrations at these times, as shown in Figure 14.5. Therefore, the starting pharmacokinetic values are changed by the fitting program slightly to see whether the predicted serum concentrations are now closer to or farther away from the actual values.

The difference between the predicted and actual serum concentration is calculated and this value is squared so that all resulting values will be positive. This entire process is performed repeatedly (i.e., by iteration) until the sum of the squares of the differences of all serum concentrations diminishes to a predefined threshold (hence, "least squares"). The program then stops the iteration and reports the last pharmacokinetic values used as the estimated patient-specific values. Figures 14.6–14.9 illustrate the sequential iterative fitting of three serum concentrations by nonlinear curve fitting. The first formula in Appendix 14.1 demonstrates nonlinear least-squares fitting.

Although iteration is little more than repeated trial and error, it is a well-established mathematical method for which many sophisticated fitting algorithms have been developed. With today's personal computers, hundreds or thousands of iterations can be performed in a fraction of a second. Nonlinear regression overcomes many of the problems of linear regression. All serum concentrations have equal weight in the fitting process. With a proper mathematical model, steady state is not required; serum concentrations over any number of dosage intervals can be fitted.

Nonlinear curve fitting can be customized in several ways. The mathematical threshold for stopping iteration can be set at different levels. The smaller the number is, the more iterations will occur before fitting is completed and the more accurate will be the estimations of the pharmacokinetic parameters. However, an extremely stringent definition is

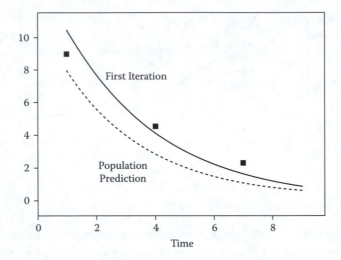

FIGURE 14.6   First iteration.

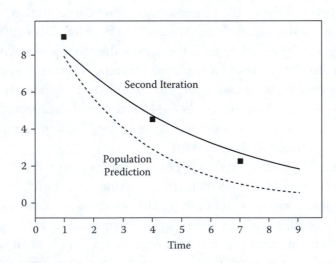

FIGURE 14.7   Second iteration.

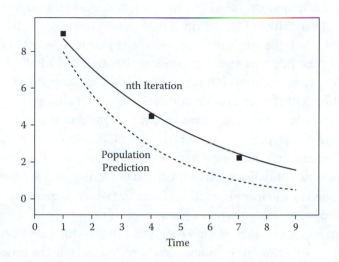
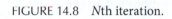

FIGURE 14.8   Nth iteration.

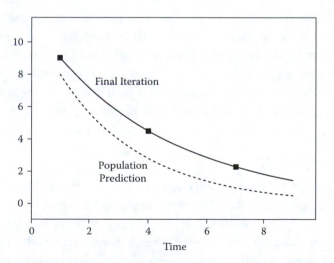

FIGURE 14.9   Final iteration.

generally not necessary for clinical use. Generation of pharmacokinetic parameters to an accuracy of two or three places after the decimal is generally adequate.

Another customization that can be implemented is systematic weighting of the serum concentration points. A common form of weighting of serum concentrations is to model the accuracy of the inherent variability of the serum concentrations themselves. In the least-squares formula in Appendix 14.1, the weighting is represented by denominators in the formula.

Laboratory test methods have some variability; although it is small, this can be incorporated into the curve fitting model. It is often modeled as a percentage of the measured serum concentration. Other sources of error include variability in the dosage administered because of inaccuracy in the measurement of intravenous doses, variability in reporting of the exact starting and stopping times of intravenous infusions, inaccuracy in recording exact times that serum concentrations were obtained (some of this variability might be minimal if bar code medication administration and blood drawing are implemented), and other variables inherent in clinical practice. These variabilities are often modeled in the equations as a fixed value rather than as a percentage of the serum concentration.

Serum concentrations can also be time weighted. That is, the curve fitting routine can be programmed to weigh older serum concentrations less heavily than more recent serum concentrations under the assumption that recent serum concentrations reflect the patient's current physiologic and pharmacokinetic state more accurately than older serum concentrations. Time weighting does reduce the overall amount of information used to estimate pharmacokinetic parameters, so the programmer needs to strike a balance between the need for modeling accuracy and information availability.

One similarity between linear and nonlinear curve fitting is that initial starting pharmacokinetic parameters are discarded in the fitting process and only serum concentrations are used in determining the final pharmacokinetic parameters. In linear fitting, *a priori* values are not considered at all. In nonlinear fitting, they gradually "fade away" during the iteration process. In the research setting in which many serum concentration values have been obtained from research subjects, the loss of *a priori* values poses no difficulties. However, in the clinical setting, financial and patient care factors limit the number of serum drug concentrations that can be obtained. Because of these limitations, a variation of nonlinear curve fitting was developed that is called Bayesian curve fitting.

14.4 BAYESIAN FITTING

The Bayesian regression method was developed to remedy some of the deficiencies of nonlinear regression in clinical practice. This methodology is based on Bayesian inference, which states that as evidence accumulates, the degree of belief in a hypothesis changes. With enough evidence, the degree of belief often becomes very high or very low.[8] In the case of clinical pharmacokinetics, the degree of belief is represented by the confidence intervals around the patient's estimated pharmacokinetic parameters. As more serum concentration data are entered into the program, the confidence in the accuracy of predicted pharmacokinetic parameters increases.

Bayesian curve fitting is a modification of the nonlinear least-squares curve fitting method (see Appendix 14.1). Applied to clinical pharmacokinetics, the method uses the

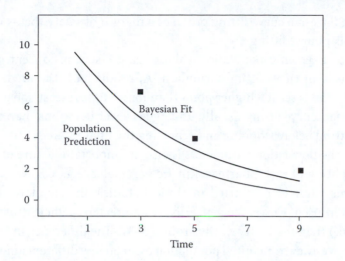

FIGURE 14.10   Final Bayesian fit.

population estimates of the patient's initial population pharmacokinetic estimates as the *a priori* values as in nonlinear curve fitting (i.e., the initial hypothesis) before. As serum concentrations are added, the resulting pharmacokinetic estimates move toward those of the nonlinear least-squares values. However, with Bayesian curve fitting, the impact of the priors diminishes much more slowly. With one or two serum concentrations, the initial population estimates can provide as much information as one additional serum concentration, thereby generating savings in cost as well as patients' blood and discomfort.[9,10]

Figure 14.10 illustrates the final iteration in the fitting of three serum concentrations by Bayesian curve fitting. In addition to the advantages of fitting points obtained at non-steady-state conditions and across dosage intervals, Bayesian curve fitting moderates the effect of unreasonable outlier serum concentrations as illustrated in Figure 14.11. Because

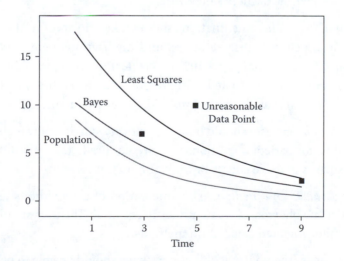

FIGURE 14.11   Outlier data point.

of this property, Bayesian curve fitting provides a margin of safety that is not provided by linear or nonlinear curve fitting.

In addition to balancing population values against serum concentrations, as illustrated in the Bayesian curve fitting formula in Appendix 14.1, the predefined threshold for stopping iteration is set at a higher point than for nonlinear least-squares curve fitting. Therefore, Bayesian curve fitting usually undergoes fewer iterations than nonlinear least-squares curve fitting before stopping and reporting the final results.[11–14]

Weighting of the population parameters is also an important feature of Bayesian curve fitting. The weighting of each pharmacokinetic parameter (e.g., *CL*, *Vd*) is related to the variability in this parameter reported in clinical studies; the standard deviation of the parameter is often used. Clearance generally has a greater coefficient of variation (typically around 50%) than the *Vd* (typically around 30%). The difference in weighting allows parameters to move away from initial population estimates by different amounts during the curve fitting—the larger the coefficient of variation, the greater the potential movement.

Bayesian curve fitting also takes into account potential mismatch between the drug's pharmacokinetics and the standard pharmacokinetic model used; this is called model misspecification. In both nonlinear and Bayesian curve fitting, the two independent pharmacokinetic variables, *CL* and *Vd*, are determined directly, rather than from *kd*, which is a dependent variable (i.e., dependent on both *CL* and *Vd* as calculated by the equation $kd = CL/Vd$).

14.5 DESIRABLE FEATURES OF A PHARMACOKINETICS COMPUTER PROGRAM

The characteristics described previously help define the features of the ideal computerized pharmacokinetic package:

- *Standard methods.* Calculation of the initial population pharmacokinetic values should use standard methods that are consistent with the prediction methods used in the clinical pharmacokinetics literature.

- *Tailored population values.* Estimations should take into account the patient's demographic data, concurrent disease states, and medications being taken that have a marked impact (e.g., a change by 10% or greater) on pharmacokinetic parameters. Parameters should be associated with reasonable variability measurements (e.g., standard deviations) that also reflect individual patient factors.

- *User modification.* The user should be able to modify starting parameters, both means and variability, for patient conditions that are not predicted by standard population estimates, such as massive edema or limb amputation.

- *Documented values and methods.* Documentation of the methods used by the program to calculate pharmacokinetic values is essential. The user should not be forced to rely on undocumented calculation methods.

- *Bayesian methods.* The pharmacokinetic package should be able to perform Bayesian curve fitting because this is currently the state of the art in clinical pharmacokinetic programs. Systems that perform only linear or nonlinear regression will suffer from the deficits itemized before.

- *Regimen calculation.* The program should use the predicted pharmacokinetic values to calculate preferred dosage regimens and serum concentrations based on a user-chosen dosage regimen.

- *Pharmacodynamic parameters.* A desirable feature is to calculate pharmacodynamic as well as pharmacokinetic end points.[15,16] For example, calculation of the area under the inhibitory curve with antimicrobial dosage regimens can help the user choose the best of several possible dosage regimens.

- *Ease of use.* It is essential that the program be easy to use. The program flow should proceed logically from demographic to population pharmacokinetics to serum concentration data to predictions to report generation.

Aside from these mandatory requirements, a number of other features are highly desirable:

- *Data storage.* Storage of past patient data in a database linked to the program will allow the user to start with stored values more tailored to the patient and to analyze new serum concentration data with previous values, making the predictions of pharmacokinetic values more robust.[17]

- *Reports.* The program should also print a report suitable for inclusion in the patient's medical record. Ideally, the program should allow the user to customize the report and provide the ability to enter free-text recommendations to the health team on management of the patient's drug therapy.

- *Graphics.* A graphical representation of serum concentration-time curves is a highly desirable feature. By looking at the graph, the user can often spot data entry errors that are difficult to discern by looking at numerical data, such as an incorrectly entered time of a dose or serum concentration. The ability to include the graph on the printed report along with target serum concentration data is also desirable.

- *Integration.* Integration of the program into a larger pharmacy system has pros and cons. Demographic and laboratory values can potentially be exported directly into the pharmacokinetics package, easing the burden of data entry for the user. However, automatic exporting of data carries risks of importing erroneous data into the program. For example, demographic data might be outdated or the laboratory may report the time at which the serum sample was analyzed rather than when it was obtained.

Appendix 14.2 shows examples from one commercially available stand-alone pharmacokinetics package for the personal computer. Listings of other pharmacokinetic programs for both clinical and research use can be found at http://www.boomer.org/pkin/soft.html.

14.6 SUMMARY

Clinical pharmacokinetics is a unique service that pharmacists can bring to the patient care setting. It can reduce the toxicity of many drugs, assist in achieving therapeutic serum concentrations more rapidly, and save costs in serum concentration sampling and hospitalization days. Computer software can aid pharmacists in analyzing serum concentration data and in standardizing practice among pharmacists at an institution.

REFERENCES

1. Bond, C. A., and Raehl, C. L. Clinical and economic outcomes of pharmacist-managed aminoglycoside or vancomycin therapy. *American Journal of Health-System Pharmacy* 2005. 62:1596–1605.
2. Welty, T. E., and Copa, A. K. Impact of vancomycin therapeutic drug monitoring on patient care. *Annals of Pharmacotherapy* 1994. 28:1335–1339.
3. Burton, M. E., Ash, C. L., Hill, D. P., Jr., Handy, T., Shepherd, M. D., and Vasko, M. R. A controlled trial of the cost benefit of computerized Bayesian aminoglycoside administration. *Clinical Pharmacology & Therapeutics* 1991. 49:685–694.
4. Bond, C. A., and Raehl, C. L. 2006 National Clinical Pharmacy Services Survey: Clinical pharmacy services, collaborative drug management, medication errors, and pharmacy technology. *Pharmacotherapy* 2008. 28:1–13.
5. Kaboli, P. J., Hoth, A. B., McClimon, B. J., and Schnipper, J. L. Clinical pharmacists and inpatient medical care: A systematic review. *Archives of Internal Medicine* 2006. 166:955–964.
6. Pedersen, C. A., Schneider, P. J., and Scheckelhoff, D. J. ASHP National Survey of Pharmacy Practice in Hospital Settings: Monitoring and patient education—2003. *American Journal of Health-System Pharmacy* 2004. 61:457–471.
7. Sanghera, N., Chan, P. Y., Khaki, Z. F., et al. Interventions of hospital pharmacists in improving drug therapy in children: A systematic literature review. *Drug Safety* 2006. 29:1031–1047.
8. Anon. Bayesian inference. Wikipedia (http://en.wikipedia.org/wiki/Bayesian_inference). Accessed August 20, 2009.
9. Burton, M. E., Chow, M. S., Platt, D. R., Day, R. B., Brater, D. C., and Vasko, M. R. Accuracy of Bayesian and Sawchuk–Zaske dosing methods for gentamicin. *Clinical Pharmacology* 1986. 5:143–149.
10. Denaro, C. P., and Ravenscroft, P. J. Comparison of Sawchuk–Zaske and Bayesian forecasting for aminoglycosides in seriously ill patients. *British Journal of Clinical Pharmacology* 1989. 28:37–44.
11. Cropp, C. D., Davis, G. A., and Ensom, M. H. Evaluation of aminoglycoside pharmacokinetics in postpartum patients using Bayesian forecasting. *Therapeutic Drug Monitoring* 1998. 20:68–72.
12. McClellan, S. D., and Farringer, J. A. Bayesian forecasting of aminoglycoside dosing requirements in obese patients: Influence of subpopulation versus general population pharmacokinetic parameters as the internal estimates. *Therapeutic Drug Monitoring* 1989. 11:431–436.
13. Radomski, K. M., Davis, G. A., and Chandler, M. H. General versus subpopulation values in Bayesian prediction of aminoglycoside pharmacokinetics in hematology-oncology patients. *American Journal of Health-System Pharmacy* 1997. 54:541–544.
14. Rodvold, K. A., Gentry, C. A., Plank, G. S., Kraus, D. M., Nickel, E., and Gross, J. R. Bayesian forecasting of serum vancomycin concentrations in neonates and infants. *Therapeutic Drug Monitoring* 1995. 17:239–246.
15. Scheetz, M. H., Hurt, K. M., Noskin, G. A., and Oliphant, C. M. Applying antimicrobial pharmacodynamics to resistant gram-negative pathogens. *American Journal of Health-System Pharmacy* 2006. 63:1346–1360.

16. Schentag, J. J. Antimicrobial action and pharmacokinetics/pharmacodynamics: The use of AUIC to improve efficacy and avoid resistance. *Journal of Chemotherapy* 1999. 11:426–439.
17. Duffull, S. B., Kirkpatrick, C. M. J., and Begg, E. J. Comparison of two Baysian approaches to dose-individualization for once-daily aminoglycoside regimens. *British Journal of Clinical Pharmacology* 1997. 43:125–135.

APPENDIX 14.1: REGRESSION FORMULAS USED IN T.D.M.S. 2000*

Nonlinear least-squares formula:

$$\sum_{j=1}^{M} \frac{(Cp_j - Cp'_j)^2}{(SD_{Cpj})^2}$$

Bayes formula:

$$\sum_{i=1}^{N} \frac{(P_i - P'_i)^2}{SD_p^2} + \sum_{j=1}^{M} \frac{(Cp_j - Cp'_j)^2}{(SD_{Cpj})^2}$$

where

N = the number of parameters fitted. In the one-compartment model, $N = 3$ for oral drugs; $N = 2$ for intravenous and intramuscular drugs;

P_i = initial (population) estimates for each pharmacokinetic parameter;

P'_i = revised (fitted) estimates for each pharmacokinetic parameter;

SD = variance of the pharmacokinetic parameter;

M = the number of serum concentrations obtained;

Cp_j = the serum concentration predicted from initial parameter estimates;

Cp_j = the predicted serum concentrations (based on revised parameter estimates);

$(SD_{Cpj})^2$ = variance of the predicted serum concentration;

$SD_{Cpj} = ([Cp_j \times SD_e] + FE) \times Q^t$;

SD_e = coefficient of variation of the assay error: Bayes: 0.1 (10%), least squares: 0.01 (1%);

FE = fixed error due to unaccounted for variability such as model misspecification; Bayes: 5% of the midpoint value of the therapeutic serum level range; least squares: 0;

Q^t = time weighting multiplier.

---

* The author participated in the development of T.D.M.S. 2000 and is a principal in Healthware, Inc.

APPENDIX 14.2: SAMPLE PHARMACOKINETICS PROGRAM SCREENS

Screen shots courtesy of Healthware, Inc. (www.tdms2000.com). Copyright 1986–2010.*
Reproduced with permission.

SCREEN 14.1   All pertinent demographic data of the patient are collected on this screen. Serum albumin is required only for phenytoin and similar drugs with binding-dependent pharmacokinetics.

SCREEN 14.2   Information on medical conditions that affect pharmacokinetic variables is collected on this screen.

---

* The author participated in the development of T.D.M.S. 2000 and is a principal in Healthware, Inc.

SCREEN 14.3 This screen shows the estimated population parameters at the top (white boxes). The user can modify the population values if desired. The user then enters desired serum concentration targets in the lower left (white boxes). In the center, the dosage regimen to achieve the desired serum concentrations is displayed (gray boxes).

SCREEN 14.4 On this screen, the user enters a trial dosage regimen on the lower left (white boxes). In the center, the resulting steady-state serum concentrations are displayed.

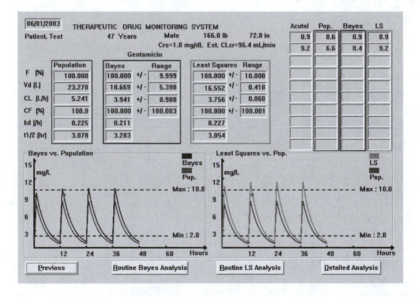

SCREEN 14.5   After the patient has received doses of the drug and serum concentrations have been measured, they are entered on this spreadsheet.

SCREEN 14.6   The program then performs Bayesian and least-squares curve fitting and displays the revised pharmacokinetic parameters and graphs the results of the dosage regimen that was entered.

SCREEN 14.7 This screen performs the same function as Screen 14.4, but this time using the revised, fitted pharmacokinetic parameters. The user can choose to print a report based on either Bayesian or least-squares parameters.

CHAPTER 15

Clinical Decision Support Systems

Pieter J. Helmons

CONTENTS

15.1 INTRODUCTION

This chapter discusses the impact of clinical decision support systems on medication errors. Therefore, it is important to understand the definitions of "adverse drug events" and "medication errors" before discussing clinical decision support systems. Adverse drug events (ADEs) are defined as any injury secondary to medication use.[1] These events can be divided into *nonpreventable, preventable,* and *potential* ADEs:

- Nonpreventable ADEs (also known as adverse drug reactions [ADRs]) are inherently associated with medication therapy. An example of a nonpreventable drug event is an allergic reaction following administration of a drug to a patient with no known drug allergies.

- Preventable ADEs are those that cause injury to the patient that could have been prevented. Using the previous example, if an allergy to the drug was known, but was ignored and the administration of the drug resulted in an allergic reaction in the patient, this would be a preventable ADE.

- A potential ADE is an ADE that could have occurred as a result of an error, but (fortunately) did not. In the preceding example, if the patient were allergic to the drug and received it, but no allergic reaction occurred, this would be a potential ADE.

Medication errors are defined as any mistakes in ordering, transcribing, dispensing, administering, or monitoring of medication.[1] This is a very broad definition and although potential and preventable ADEs are medication errors, not all medication errors are ADEs.

Both medication errors and ADEs are common, costly, and cause clinically important problems.[2,3] Each year, an estimated 770,000 people are injured or die in hospitals from ADEs. Approximately 28% of adverse drug events are the result of medication errors and are therefore preventable. More than half of these medication errors occur at the drug ordering stage and are the result of insufficient patient-specific information at the time of prescribing[1,4] (see Chapter 10).

This chapter starts with a case that illustrates how medication errors can result from the lack of patient-specific information. Next, the same case is presented, but this time the healthcare provider is supported by a clinical decision support system, resulting in an entirely different scenario and patient outcome. Although this alternative scenario lacks a specific pharmacist intervention, the crucial role pharmacists play in designing and maintaining these systems will be discussed later in this chapter.

15.1.1 Case before Clinical Decision Support[5]

Patient X is a 62-year-old woman with diabetes, hypertension, and borderline kidney failure. She has been seeing her primary care physician, Dr. Smith, for the past 3 years and has generally been pleased with her care. She arrives at the office for a visit, checks in at the front desk, and then is ushered into an examination room. A few minutes later, Dr. Smith enters the room to see her. He is carrying her paper chart, and he flips through it as they discuss her current issues. After some discussion and a brief physical examination, Dr.

Smith determines that patient X has a sinus infection. He glances at the medicines she is taking and his last written note about drug allergies, and then he hand-writes a prescription for an antibiotic. Patient X leaves the office with the written prescription and takes it to her pharmacy. The pharmacist enters the prescription into his computer system and then informs patient X that the antibiotic is not covered on her benefit plan. The pharmacist places a call to Dr. Smith's office, resulting in the prescription of an alternative antibiotic.

Patient X receives the antibiotic and instructions from the pharmacist about how to take the drug and then returns home. That evening she takes the first dose of the drug; an hour later, she develops severe vomiting. Patient X calls her doctor's office to report the new problem. When the message reaches Dr. Smith, he considers that perhaps the drug was given in too high a dose given her age and kidney function. He lowers the dose of the antibiotic and prescribes an antinausea medicine. The antinausea medicine eventually controls her vomiting but makes her very sleepy—so much so that when she gets up that evening to go to the bathroom, she stumbles and falls, breaking her hip. She is taken to the hospital by ambulance and undergoes surgery the next morning to have her hip stabilized with pins.

15.1.2 Case after Clinical Decision Support

Patient X arrives for her office visit. The nurse brings her to the examination room and puts a preliminary diagnosis of "sinus infection" into the computer. Dr. Smith arrives to see her a few minutes later. After examining her and confirming the preliminary diagnosis, Dr. Smith clicks a button to reveal an evidence-based recommendation on the best antibiotic options for this condition. The computer returns a list of three antibiotic choices; next to each choice is an icon indicating whether that medication is covered on patient X's plan. The first antibiotic is nonformulary, so Dr. Smith selects the second antibiotic. The computer checks the patient's other active medications, and an alert window pops up indicating that the drug may interact with one of her diabetes drugs, resulting in vomiting. (In fact, it was this interaction, not the patient's age or kidney function, that was responsible for her vomiting in the first scenario; in that scenario, the physician did not make this connection.)

Dr. Smith contemplates giving patient X a reduced dosage of the drug and treating despite the risk of vomiting. To be sure, though, he clicks a button revealing her drug history over the past 3 years. He notes that one of his partners gave a similar drug to her last year and the result was, indeed, severe nausea and vomiting. Armed with this highly relevant history, Dr. Smith cancels the drug order and selects the third antibiotic. No warnings appear this time, but the computer does recommend a reduced dosage based on her age and last measured kidney function, which Dr. Smith accepts. He confirms the prescription with a click, which directs the prescription to be electronically transmitted to the patient's local pharmacy, and also prints a concise patient's guide to the drug and its potential side effects. He reviews the prescription, dosage, and potential side effects with patient X and prepares to discharge her from the office.

Before sending her home, however, he notes that the computer, which includes a full electronic health record as well as an electronic prescribing function, is recommending that the patient be placed on a cholesterol-lowering drug, based on her most recent cholesterol and LDL results and her diagnosis of diabetes; the system again shows which of the applicable drugs is on the formulary of the patient's plan. With two clicks, Dr. Smith prescribes this medication as well—again following the computer's recommended adjustment for age and kidney function. The computer also recommends a follow-up blood test (creatine kinase) after 4 weeks of therapy because of the potential risk of muscle inflammation with this family of drugs. With one click, Dr. Smith orders this blood test and instructs the patient to return in 4 weeks to get the test done. The rest of patient X's course remains uneventful and she recovers rapidly from her sinus infection without further incident.

15.2 INTRODUCTION TO DATA, INFORMATION, KNOWLEDGE, AND DECISION SUPPORT

15.2.1 Definitions

In pharmacy informatics, the words data(base) and knowledge (base) are often used. To better understand the definition and function of decision support systems, it is essential to understand the difference between these terms (Figure 15.1). A datum (the word "data" is plural) is defined as a single observation that characterizes a relationship; in other words, it is the value of a specific parameter for a specific object (e.g., a patient).[6] Knowledge is derived from the formal or informal analysis of data. As an example, if the result of a single measurement of a patient's blood pressure is 180/110 mm Hg, this is considered a datum. An analysis of a large number of blood pressure measurements in a population leads to the reference values of normal, high, and low blood pressures. This analysis has now resulted in knowledge on patient blood pressure.

A database is a collection of individual observations without any summarizing analysis. A computerized medication record is primarily a database; only data on the patient's medication are stored. However, if (medical) knowledge is added to these systems (e.g., reference values of kidney function or knowledge of interactions between medications), the computer may apply this knowledge to aid in case-based problem solving. The system is then a *knowledge-based* system or *decision support* system.

This brings us to the definition of a clinical decision support system (CDSS)[7]: "software that is designed to be a direct aid to clinical decision-making, in which the characteristics of an individual patient are matched to a computerized clinical knowledge base and patient-specific assessments or recommendations are then presented to the clinician or the patient for a decision." These systems convert patient *data* essential for the clinician to make the right decisions into usable *information* at the time of decision making.

Typically, a CDSS is based on the following elements (see Figure 15.1):

- The *knowledge base* translates scientific knowledge (e.g., guidelines, treatment protocols) into computer-interpretable decision algorithms (e.g., clinical rules or algorithms)

- The *rules engine* retrieves patient-specific data, often stored in multiple databases, and checks whether the criteria set in the knowledge base are met.

- *Software* allows the user to create clinical decision algorithms and generates recommendations.

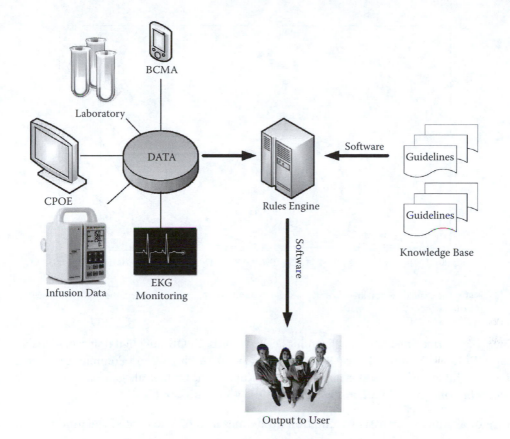

FIGURE 15.1 Elements of a CDSS. Clinical guidelines (knowledge base) are translated to computer interpretable decision algorithms (clinical rules). The rules engine is then used to match patient-specific information to the parameters specified in the clinical rule (e.g., the current dose of a medication is matched to the renal function of the patient). If dosage adjustment is warranted according to the criteria in the knowledge base, the user is notified. CPOE: computerized provider order entry; EKG: electrocardiogram; BCMA: bar-code-enabled medication administration.

15.2.2 Why Are Decision Support Systems Needed?

The Institute of Medicine report, "Crossing the Quality Chasm," has documented the gap between what healthcare providers know and what they do.[8] The report identified three types of quality problems: overuse, underuse, and misuse. Misuse (errors) has been the predominant focus of attention, but it is likely that underuse or overuse of practices and resources results in a larger portion of current quality problems.[9]

Surveys of clinicians indicate that a major barrier to using current research evidence is the time, effort, and skills needed to access the right information among the massive volumes of research.[10] Each year, the National Library of Medicine indexes over 560,000 new scientific articles in the MEDLINE database. In addition, 20,000 new randomized trials are added to the Cochrane Library.[9] This corresponds to 1,500 articles and 55 new trials per day! Even if the clinician is aware of the evidence, he or she needs to agree, adopt, and adhere to this evidence. For example, in one study, 90% of the clinicians were

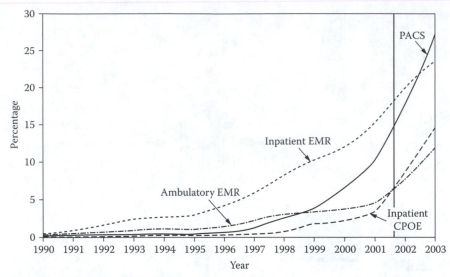

Note: The shaded vertical line illustrates a suggested shift for the curves to reflect the "have it in place" measure of adoption.

FIGURE 15.2   Implementation of electronic health records, CPOE, and digital storage of diagnostic images; EMR = Electronic Medical Record; PACS= Picture Archiving and Communication Systems (digital archiving of diagnostic images). (From Kuperman, G. J. et al. *Journal of the American Medical Information Association* 1998. 5:112–119. With permission.)

aware of acellular pertussis vaccination guidelines and 67% accepted the guideline; however, only 35% adhered to the guideline.[11] In addition, patient acceptance of and adherence to treatment plans are often problematic. If 80% adherence to each of these stages were to be achieved, this would still result in evidence-based treatment of only 21% of the eligible patient population ($0.8^7 = 0.21$).[10]

Decision support systems can only be as good as the data on which the system is based. The 1999 Institute of Medicine report, *To Err Is Human*, has resulted in an enormous focus on medical and medication errors. Some of the conclusions of this report were that errors were often the result of poorly designed systems and that healthcare facilities should rely more on automation to make the system less error prone.[12] As a result, most hospitals have implemented or are implementing hospital information systems (see Figure 15.2[13] and Chapter 6), most hospital pharmacies have implemented pharmacy information systems (see Chapter 7), and most hospital nursing units use automated dispensing cabinets, limiting the access to medications. In addition, medication administration errors are being addressed with bar-coded medication administration (see Chapter 8) and intelligent ("smart") infusion pump technologies.

During the last decade, computerization has led to an exponential increase of patient-specific data that can be used in decision-support algorithms. In the near future, the field of genomic medicine will provide patient-specific genomic data that can be incorporated into algorithms. Already, decision support is considered essential to integrate the vast amounts of genomic data with "traditional" parameters.[14] Some experts estimate that in just a few years primary care physicians will have to know how to employ as many as 100,000 new genetic screening tests,[15] further stressing the important role of decision support.

The focus on quality of care and the increased availability of electronic data have resulted in greater performance requirements for healthcare organizations. The Joint Commission has implemented standardized performance measures that are designed to track the performance of hospitals and encourage improvement in the quality of care. These indicators are derived from current consensus guidelines and represent current standards of care. As an example, one performance indicator measures the percentage of patients eligible for pneumococcal vaccination that were actually vaccinated while admitted to the hospital. In 2002, 28% of patients were vaccinated; this improved to 50% in 2004.[16] Decision support could be used to inform clinicians of these performance indicators, select eligible patients, and further improve adherence to guidelines.

Most pharmacy information systems currently provide some degree of basic decision support intended to support the pharmacist in the evaluation of the patient's medication profile. Recently, 30 clinical pharmacy information systems were tested to see whether they could prevent 18 unsafe medication orders. These orders had been selected because they had already caused severe adverse outcomes in patients.[17] Only 67% of these systems were directly interfaced with the laboratory system, which is essential for drug–laboratory interaction checking.

This study showed that, on average, only 44% of the unsafe orders were detected by these systems. Also, 50% of these systems routinely generated recommendations that were of little to no clinical value. Decision support could improve the performance of these systems by integrating additional patient-specific information, resulting in more clinically relevant recommendations.[5]

15.3 USING DECISION SUPPORT SYSTEMS TO IMPROVE PHARMACOTHERAPY QUALITY

Decision support systems have been used to guide clinicians to the most likely diagnosis, to remind clinicians of measures to prevent disease (e.g., pneumococcal vaccination), to improve the management of disease (e.g., improving diabetes care by preventing complications), and to improve appropriate selection, dosage, and monitoring of drug therapy (Figure 15.3). This section focuses on this last category because most pharmacists will be involved in decision support as part of pharmacy information systems or computerized provider order entry (CPOE) systems. These systems can be categorized as basic CDSS and advanced CDSS.

15.3.1 Basic Clinical Decision Support

Drug allergy checking presents an alert when a clinician orders a medication to which the patient has an electronically documented allergy. Most pharmacy systems have this functionality because it is considered an important patient safety feature. However, these systems are often far from perfect.[18] Major shortcomings include:

- There is no requirement for structured, coded entry of allergens (i.e., a controlled vocabulary; see discussion of standards and controlled vocabularies in Chapter 4). This makes it impossible to be alerted to cross-reacting allergens within the same drug class and to transfer allergy information between information systems.

- If allergy data are coded, cross-reactivity data do not distinguish between a theoretical cross-reactivity and an evidence-based contraindication.

- Allergy data in the database can be of poor quality. A recorded allergy is often considered a definite contraindication for the patient, sometimes resulting in withholding the most appropriate therapy. However, the documented allergy can be based on a side effect (e.g., diarrhea from antibiotics) or a mild allergic reaction (e.g., a minor

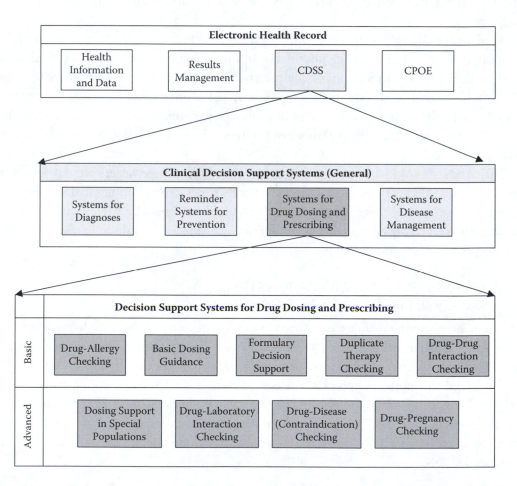

FIGURE 15.3 The spectrum of decision support. (Kuperman, G.J. et al. *Journal of the American Medical Information Association* 2007. 14:29–40; Blumenthal, D., and Glaser, J. P. *New England Journal of Medicine* 2007. 356:2527–2534; Garg, A. X. *JAMA* 2005. 293:1223–1238.) The first bar depicts clinical decision support systems (CDSS) as part of a patient's electronic health record. The second bar depicts the different applications of CDSS in healthcare. The third bar shows the application of CDSS in pharmacy information systems.

rash from an antibiotic). Also, allergy data of a patient are seldom updated. Once an irrelevant allergy is recorded, physicians are very reluctant to delete this warning.

These shortcomings and the rare occurrence of a definite allergy in the general patient population have led to excessive, irrelevant drug-allergy alerting.

In nonautomated ordering environments, dosage errors are the most common type of medication error leading to preventable ADEs.[1] Susceptible patients, such as children and the elderly, are at risk of serious dosage errors, especially overdosage.[19–21] Even basic decision support within CPOE can dramatically improve appropriate dosage of medication by:

- providing the clinician a list of patient-specific dosage parameters (often based on the age of the patient);

- drug-specific dosage parameters (based on predefined minimum and maximum allowed dosages); and

- indication-specific dosage parameters (the prescriber selects the indication of a specific drug and drug dosages are automatically entered based on the selected indication).

Eliminating manual dosage entry also decreases the potential for a wrong decimal point, typographical error, or wrong dosage unit (e.g., milligrams instead of micrograms) in the medication order. However, apart from the patient's age, basic dosage guidance often does not take into account other patient-specific parameters, such as renal function and electrolyte levels.

A classic example is the following: A physician prescribes a normal dosage of an antibiotic for a 45-year-old patient with renal failure. No dosage alerts are generated because the patient's renal function is not used to provide dosage recommendations. In fact, had the physician adjusted the dosage appropriately in this patient, he might have been alerted to prescribing a subtherapeutic dosage. Thus, if an error is made, no alert is generated, but if the physician prescribes the appropriate dosage, an irrelevant alert is generated.

Most hospitals try to control the rising costs of drugs by maintaining a formulary: a selection of drugs covering all therapeutic areas that can be used in the hospital. This selection is based on providing essential medications to support safe and effective care, while preventing or limiting the use of high-cost drugs with limited additional benefit. Basic decision support can improve formulary compliance by assisting clinicians in the selection of formulary options over nonformulary options. One approach is to display a pop-up alert when the clinician attempts to order a nonformulary drug, while at the same time providing a selectable list of similar formulary medications. This approach can be very successful if alerts include clear and to-the-point guidelines with links to additional information and if noncontroversial alternatives are suggested within the same alert window (see Figure 15.4).[16,22]

Duplicate therapy occurs when more than one regimen of a single drug or multiple regimens of different medications with similar therapeutic effects are prescribed. It often occurs in situations in which several clinicians provide care for the same patient. Duplicate

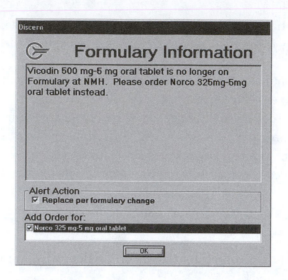

FIGURE 15.4   Formulary alert with one-click correction capability. (From Kuperman, G. J. et al. *Journal of the American Medical Information Association* 1998. 5:112–119. With permission.)

orders also originate from switching from intravenous therapy to oral therapy with the same drug without discontinuing the original intravenous order. Therapeutic duplication is uncommon (less than 6% of all prescribing errors are duplicate orders[1]), but often results in a large number of irrelevant alerts. Prescribing multiple drugs from the same drug class is very common (and appropriate) for antimicrobials, immunosuppressants, opioids, and insulin. Also, when dosage tapering occurs and different doses for the same drug are ordered, intentional duplicate orders exist in the patient's medication profile.

The relatively rare occurrence of unintentional duplicate orders and the large number of irrelevant alerts resulting from basic CDSS have caused organizations to inactivate duplicate alerting altogether.[23] Extensive customization of duplicate order checking and selective alerting are needed to prevent excessive irrelevant alerting. Examples of successful customization are limiting duplicate order checking to classes with high risks of adverse events (e.g., analgesic, cardiac, psychiatric, and endocrine medications).[23] Also, increasing the number of relevant alerts by further customizing the alert logic is essential to prevent desensitization to all classes of alerts.

Computerized drug–drug interaction checking is one of the most frequently used types of CDSS. However, as with duplicate order checking, drug–drug interaction checking is associated with large numbers of clinically unimportant alerts. In one study, 11% of all medication orders generated a drug–drug interaction warning and clinicians overrode 88% of the interactions that the system considered a "critical" drug–drug interaction.[24] Also, clinicians categorized only one in nine interactions as potentially relevant at the time of the warning.[25] Another study found that adverse consequences almost never occurred, even when the highest level of drug–drug interactions was overridden.[26] However, a number of clinically relevant interactions are likely to go unnoticed and lead to adverse patient outcomes because their alerts are buried in a mountain of irrelevant alerts. The most important reasons for this large number of irrelevant alerts include[18]:

- Vendor-supplied drug interaction knowledge bases have no or limited flexibility for modifications (i.e., only allow the display of the most relevant interactions).

- Flawed logic triggers the alert. Patient-specific parameters needed to generate a clinically relevant alert are not included in the clinical rule, leading to irrelevant alerts. An example of this is the hyperkalemia warning when spironolactone (an aldosterone receptor antagonist) is prescribed together with an angiotensin converting enzyme inhibitor. This is a very common combination in patients with heart failure and leads to a large number of alerts because the actual potassium level of the patient is not integrated into the clinical rule. Ideally, an alert should appear only if the patient already had a high or high-normal serum potassium level and the aforementioned drug combination was prescribed.

- There is no discrimination between the presentation of a highly clinically relevant interaction that warrants immediate action and an interaction of minor importance. A similar presentation of a serious alert (e.g., a definite allergy to penicillin) and a minor alert (e.g., "draw potassium levels within the next 3 days") could lead to an override of both alerts. A recent study showed that discrimination between alerts leads to a higher acceptance rate of serious alerts by clinicians.[27]

The value of basic decision support could dramatically increase if these limitations were addressed.

15.3.2 Advanced Clinical Decision Support

Implementing decision support in a complex healthcare environment is a daunting task. It is therefore recommended that *advanced* medication-related decision support should be implemented only after basic decision support is in place and working well, with good user acceptance.[18] However, most studies showing important safety and financial benefits of decision support have focused on the evaluation of advanced clinical decision support.

As noted earlier, basic clinical decision support systems sometimes assume that patients are nongeriatric adults with normal physiologic function. However, to determine accurately what is a safe and appropriate dosage for a particular patient may require many factors to be considered. Some of these factors are age, weight, and height of the patient; the indication for the drug; renal function; liver function; fluid status; concomitant medications; genetic predisposition; and reactions to previous medications.

Each of these conditions affects large patient populations; in one study, 42% of inpatients had some degree of renal insufficiency.[28] Although these parameters are not always relevant for all drugs, advanced decision support can integrate these parameters for dosage recommendations in relevant cases. In one example of advanced dosage support,[29] a CDSS generated dosage recommendations of antibiotics based on the patient's age, renal function, and the sensitivity pattern of the infecting microorganism. This program substantially decreased the number of adverse events, number of days of unnecessary therapy, and costs.

Several categories of drugs need monitoring of their serum concentration (e.g., aminoglycoside antibiotics, digoxin, and antiepileptic drugs) or of the physiological parameter affected (e.g., the prothrombin time with coumarin derivatives such as warfarin). Decision support tools remind physicians to request the appropriate blood samples at the appropriate time. In one study, the number of antiepileptic blood levels that were drawn inappropriately decreased from 54 to 14.6% after implementing a decision support system.[30] Another study showed that alerts at the time of ordering could double physicians' rates of compliance with a variety of guidelines, including drug monitoring.[31]

Integrating laboratory values with drug–drug interaction checking can greatly decrease the number of irrelevant alerts. However, access to the patient's previous laboratory results is an important prerequisite of medication–laboratory test monitoring. But even when laboratory values are incorporated into the decision support system, rigorous evidence on monitoring is often lacking. Most recommendations are currently based on expert opinions or package inserts that are often nonspecific (e.g., "periodic laboratory testing is recommended"), complicating the development of explicit decision support rules.[32]

Clinicians should avoid prescribing contraindicated drugs based on preexisting disease states and other patient-related conditions. A review of drugs in the British National Formulary revealed around 1,500 contraindications between drugs or drug classes and morbidities or clinical states.[33] The most important contraindications are renal impairment and hepatic impairment. Accurate medication-contraindication checking has been a daunting task for several reasons. Similarly to other categories of decision support, the information about contraindications for healthcare providers is often vague and unstructured. For example, streptokinase, an agent used in dissolving blood clots, is contraindicated in "all conditions that are likely to be associated with existing or very recent hemorrhage"[34]—without defining likelihood, which conditions, or what constitutes "very recent."

Second, contraindication decision support only works when patients' diagnoses and conditions have been accurately entered as structured data into the electronic health records. However, the diagnosis for a patient's admission is often not entered until the patient is discharged. Contraindications related to hepatic or renal impairment are often dependent on the degree of impairment: Different alerts should be presented if a patient has severe renal failure, as opposed to moderate renal failure. Finally, no simple test is available to rate liver function impairment in a fashion similar to that for renal function impairment.

Drug–pregnancy alerting is an important category of advanced decision support. A small number of drugs should never be prescribed to a woman who is or might be pregnant (e.g., thalidomide, isotretinoin). Even if drug–pregnancy interactions were appropriately classified, the biggest challenge in this category of decision support would still be to determine the pregnancy status of the patient accurately. Pregnancy tests are not routinely performed upon admission and many systems do not contain the results of recent pregnancy tests. Also, some systems do not update pregnancy information when the pregnancy has ended.

Not surprisingly, this category of decision support also suffers from a large number of irrelevant alerts; in one study, only 10% of the drug-pregnancy alerts led to a cancellation of the offending drug, and 90% of the alerts were ignored![23] Thus, in order to benefit fully from the categorization of drugs in pregnancy, electronic health records should allow clinicians to document the pregnancy status explicitly (is pregnant, might be pregnant, etc.).

15.4 DEVELOPMENT OF CLINICAL DECISION RULES AND PROTOCOLS

15.4.1 Paper Protocols, Clinical (Decision) Rules, and Computerized Protocols (Algorithms)

Clinical care is determined by clinicians' decisions and by each patient's individualized expression of his or her illness. However, most paper guidelines are far from individualized and lack specific instructions for many of the scenarios encountered in clinical practice.[35] If patient-specific parameters are not considered in medical decision making, legitimate concerns are raised about patient-invariant ("cookbook") care.

CDSSs by themselves also contain different levels of individualized decision support. The basic level of decision support is generated through *clinical rules*. These rules have a typical "IF, THEN" logic: *If* a patient meets a standardized set of criteria, *then* an alert is generated. Basic clinical rules are very useful for "simple" drug–laboratory interactions, but fall short when decision support systems are used based on complex treatment guidelines. This is when *computerized protocols* are very useful. Computerized protocols are similar in structure to the decision flowcharts commonly used in paper guidelines,[36] but they contain much more detail than paper guidelines and clinical rules. Computerized protocols are a combination of multiple clinical rules.

Decision support systems standardize clinical decisions for patients. This is not synonymous with "each patient receives the same treatment." For example, a clinical rule can be created standardizing the monitoring of patients receiving thiazide diuretics known to decrease serum potassium levels. The clinical rule takes current serum potassium levels and co-medication into account. In a patient with a low potassium level, the same rule will recommend addition of a potassium-sparing diuretic or potassium supplementation; in another patient with a physiological potassium level, no recommendation is generated. The clinical rule is identical but the outcome is different. This is very important because these clinical rules are now generic and, if proven effective, can be used by other hospitals.

15.4.2 Stages in Clinical Decision Rules and Computerized Protocol Development[10,37,38]

Clinical decision rules and computerized protocols are designed to help clinicians with diagnostic and therapeutic decisions. These tools help clinicians cope with the uncertainty of medical decision making and help clinicians improve their efficiency. Because computerized protocols consist of multiple individual clinical decision rules, the essential steps in the development of individual clinical decision rules are also applicable to computerized protocols. Creating clinically relevant and effective clinical decision rules follows the six steps summarized in Table 15.1.

The obvious first step is assessment of the need for a decision rule. An organization should ask itself: Is there a variation in clinical practice resulting in suboptimal patient therapy? How often are clinicians currently not adhering to established (paper) treatment

TABLE 15.1 Six Steps in the Development of a Clinical Decision Rule

Stage	Factors
1. Is there a need for the decision rule?	Prevalence of the clinical condition in the hospital's patient population
	Variation in practice leading to decreased quality of care
2. Was the rule derived according to methodological standards?	Selection of subjects
	Definition of outcome
3. Has the rule been prospectively validated and refined?	Accuracy of the recommendations
	Completeness of rules: Does the tool accommodate most clinical circumstances?
4. Has the rule been previously successfully implemented into clinical practice?	Effects that can be expected from implementing the clinical rule (if known)
	Acceptance of the rule by clinicians
5. Would implementation of the rule be cost-effective?	Is cost saving a goal of the decision rule?
6. How will the rule be disseminated and implemented?	Selection of the appropriate care area
	Type of alert that is generated (obtrusive, unobtrusive)

Source: From Stiell, I. G., and Wells, G. A. 1999. *Annals of Emergency Medicine* 33(4):437–447.

protocols that could decrease this variation? Can decision support be applied to tackle this problem? If the answer is yes, then the second step is a thorough evaluation of the (paper) treatment protocols. There may be valid reasons for not adhering to a certain protocol, such as a different patient population and comorbidities. This is why thorough evaluation (and refinement, if necessary) of the decision rule should occur prior to implementing the rule in clinical practice (step three). The next section will discuss this step in more detail.

The fourth step is to investigate the effects of a similar decision rule in other organizations. What were the effects? How did the rule perform? The fifth step is the requirement of the clinical rule to be cost effective. This is applicable to situations where clinical rules are developed to increase efficiency. It should be emphasized that not every clinical rule saves money!

In fact, better adherence to treatment guidelines can initially generate more costs for an individual hospital, but ultimately lead to better patient outcome and decreased costs for society as a whole. An example is the requirement to treat every patient who suffered from a myocardial infarction with a beta-blocker. Increasing adherence to this guideline from 75 to 90% will initially lead to higher beta-blocker use expenses for the hospital. However, it will ultimately lead to fewer secondary myocardial infarctions and future hospitalizations.

The final step is to evaluate the best way to implement the clinical decision rule. Is this rule applicable to the whole hospital (e.g., dosage adjustment of antibiotics in renal function impairment) or only specific care areas (e.g., clinical rules developed to assist in the prescription or administration of oncology medication)? Is it necessary to generate an instant, obtrusive alert in the electronic prescribing system when the rule is triggered or is a weekly reminder by e-mail or page sufficient?

Involvement of all relevant clinicians (physicians, nurses, pharmacists) in all six stages of the process is critical for success.

15.4.3 Validating and Refining Rules: Positive Predictive Value as a Performance Indicator

A clinical decision rule or a computerized protocol is designed to improve the quality of care. It is therefore essential to validate the output prior to implementation of the rule or protocol in clinical practice. This is especially important and challenging for computerized protocols because they consist of many individual decision rules with many outputs. Constant monitoring of the performance of computerized protocols is also recommended after implementation in clinical practice.

A commonly used parameter to monitor performance of a CDSS is the positive predictive value (PPV).[39] PPV is defined as the number of clinically appropriate recommendations generated by the CDSS divided by the total number of recommendations generated. Ideally, the PPV should always be 1 (or 100%) because that means the recommendations generated by the system are always appropriate. In practice, this maximum PPV is seldom obtained for several reasons: A maximum score would mean that the required data in the patient's electronic medical record is always available and correct and that the computerized protocol is always applicable to all patients. However, depending on the rule, PPVs of 80–90% are possible.[39,40] Compared to conventional drug–drug interaction checking with PPVs of about 30%,[27,39,41] these PPV values are an enormous improvement.

15.5 BARRIERS TO IMPLEMENTATION[5]

Although the potential of CDSS is clear, very few hospitals and other healthcare institutions have implemented a CDSS. Moreover, the necessary electronic infrastructure needed to implement a CDSS is absent in almost 20% of U.S. hospitals.[42] The 2008 CDSS and electronic medical record statistics are depicted in Table 15.2. This table shows the very low number of hospitals that have implemented advanced CDSS (stage 4 and higher), indicating that even hospitals that capture essential patient data electronically have not achieved the next step of using these data in a CDSS. This section focuses on the barriers associated with these low adoption rates.

15.5.1 Lack of Standards and "Reinventing the Wheel"

Table 15.3 lists 10 of the most common barriers impeding widespread use of a CDSS. It is because of these barriers that the implementation (and the published research) of advanced CDSS is largely limited to four benchmark research institutions.[43] Barriers 5 and 6, "local management of the knowledge base" and "lack of standards for patient data," especially make widespread implementation and sharing of clinical rules and guidelines almost impossible.

This has a number of important implications for institutions implementing decision support. In order to be commercially viable, commercial clinical decision support systems rely on limited patient data available in most hospitals (medication and laboratory data), making advanced decision support through computerized protocols impossible. As

TABLE 15.2   Electronic Medical Record (EMR) Adoption, 2008

Stage	Cumulative capabilities of EMR[a]	2007 (%)	2008 (%)
Stage 7	Medical record fully electronic; health care organization able to contribute clinical care data as by-product of EMR; data warehousing in use	0.0	0.3
Stage 6	Physician documentation (via structured templates); full CDSS; full PACS*	0.3	0.5
Stage 5	Closed-loop medication administration (tightly coupled hospital and pharmacy systems integrated with bar coding technology at the patient's bedside	1.9	2.5
Stage 4	CPOE and advanced CDSS implemented (clinical protocols)	2.2	2.5
Stage 3	Clinical documentation (via paper flow sheets); CDSS (basis error checking); PACS data available outside of radiology	25.1	35.7
Stage 2	Clinical data available in electronic format allow physician access to review and retrieve patients' results	37.2	31.4
Stage 1	All three ancillary major hospital data systems (pharmacy, laboratory, and radiology) are installed	14.0	11.5
Stage 0	Some clinical automation may be present, but all three of the major ancillary systems (pharmacy, laboratory, radiology) are not installed	19.3	15.6
	Total hospitals surveyed	$n = 5,073$	$n = 5,466$

Source: From HIMMS Analytics. EMR Adoption Model 2008. http://www.himssanalytics.org/hc_providers/emr_adoption.asp (accessed March 26, 2009).

[a] Each stage includes the capabilities of the previous stage.

TABLE 15.3   Barriers to Widespread Adoption of CDSS

Barriers

1. Limited CDSS capabilities of existing CPOE products
2. Limited usability of systems and CDSS modules
3. Limited access to patient data needed to support a CDSS
4. Limited access to best CDSS knowledge
5. Local management and maintenance of the CDSS knowledge base
6. Lack of standards for data, medication dictionaries, cost calculations, etc.
7. High cost and difficulty of implementation
8. High cost of use and maintenance
9. Difficulty in recognizing and objectifying value
10. Perception of increased liability if CDSS recommendations are rejected

Source: From Teich, J. M. et al. *Journal of the American Medical Information Association* 2005. 12(4):403–409.

a result, advanced decision support guidelines that are effective in one institution cannot be readily implemented in other institutions. This "reinventing of the wheel" not only impedes CDSS implementation, but also is very costly.

15.5.2 Concerns about Quality and Safety Aspects of CDSS

An important barrier is the fear of decreased alertness of clinicians toward systems recommendations ("the computer is always right" situation). This phenomenon is described in the literature and has led to severe patient harm in different areas.[44,45] Simply acting on systems' recommendations without considering the full clinical picture is dangerous and also likely to occur. This is why clinical decision rules and protocols should be thoroughly validated. Also, the systems should clearly communicate to the clinician that certain areas are not covered by a specific decision algorithm. Further research is needed to minimize the risk of these unintended consequences.[4]

Most of the decision support modules are part of a CPOE system (see Figure 15.3). Very few systems can be purchased as add-ons to existing systems.[18] System developers and vendors should be clearer about the limitations of their technologies. Often, more is expected from a system than the system can deliver. Commercial systems are often designed with a "one size fits all" philosophy. Although probably more commercially viable, these systems are not designed to be integrated into the user's work flow and often do not provide the flexibility that is needed to better fit real-world clinical practice.[46]

15.5.3 Gaining Acceptance by Healthcare Professionals[38]

Often, an alert is intended to do more than transfer information. Alerting is about generating effect: The developers of the rule want to make sure that clinicians will act on their recommendations.[46] Current (basic) decision support systems all suffer from the same problem: They often trigger irrelevant reminders and alerts. It is no surprise that in a situation where time is a scarce resource and too many of the alerts are either irrelevant or overly predictable, irritated pharmacists and physicians disregard relevant and irrelevant alerts altogether. This is called "alert fatigue" and can be prevented in several ways[46]:

- Develop only clinically relevant rules and algorithms; develop decision support algorithms only if the current situation is not optimal and if preliminary research has shown that decision support can improve the situation. The results from this preliminary research should be communicated to the clinicians.

- Validate and monitor the performance of the clinical rule. Present highly clinically relevant warnings as readily identifiable and easily distinguished from other warnings.[18] An example is to have a daily e-mail sent to draw blood samples for drug concentration measurements based on standard pharmacokinetic advice, but to have an obtrusive alert pop up or a page sent out instantly when a drug concentration is potentially toxic.[4]

- Develop strategies to integrate recommendations in clinical work flow. In general, from a human factors standpoint, there is a lack of knowledge about the best way to present specific types of alerts to providers. An example is formulary management decision support. Formulary adherence greatly improved when clinicians were provided with real-time alerts that included a link to an alternative and additional information as needed (FDA alerts, drug shortages, etc.) (Figure 15.4).[18] In fact, automatically providing formulary decision support to the user was the most important determinant of improved clinical practice by a CDSS.[47]

An important barrier to healthcare professionals' acceptance of CDSS is the perception of increased liability if the recommendations provided by the system are rejected. Again, acceptance can be increased by thoroughly validating decision support algorithms prior to implementation and always allowing the clinician a "way out." The reasons why an alert or recommendation was not followed should be captured by allowing the clinician to enter a reason. This important information not only is an essential part in the continuous performance improvement of the decision algorithm, but also serves as documentation of the clinician's decision. In addition, it is proposed that clearly stated liability considerations and appropriate liability protections should be developed and clinicians educated about this subject.[5]

Although very few studies specifically address the cost of developing and implementing CDSS, there is no doubt that these systems are very costly. The price of a basic (out-of-the-box) CDSS starts around $30,000. However, due to the lack of universal standards, developing and validating the clinical rules and algorithms can cost millions.[48] Clearly, the cost-benefit ratio of these systems depends on the quality-of-care issues they intend to improve. Further research is warranted to identify interventions that are most cost effective in direct costs (hardware and software) as well as indirect costs (manpower and maintenance).

15.6 RECOMMENDATIONS AND FUTURE AREAS FOR RESEARCH[18]

Recently, a road map for national action on clinical decision support has been developed to take away the barriers mentioned in the previous sections and to improve national adoption of this potentially powerful technology.[15] The road map identifies three pillars that need to be in place to benefit fully from the potential of CDSS:

1. Best knowledge available when needed in standard formats. The best available clinical knowledge is well organized, accessible to all, and written, stored, and transmitted in a format that makes it easy to build and deploy CDSS interventions that integrate the knowledge into the decision-making process. Assuring adequate informatics education among clinicians is essential to reach this goal. These clinical informaticians are needed to bridge the gap between clinical and technological worlds because they speak the language of both and therefore can act as translators[46] (see Chapter 9).

2. High adoption and effective use (high compliance). CDSS tools are widely implemented and extensively used. Only wide national implementation of a CDSS fully exploits the potential of this technology. This means that incentives (usually financial) need to be created for organizations to implement CDSS and for benchmark

institutions to share their knowledge. Also, further research is needed to optimize alerting methods and to prevent alert fatigue.[18]

3. Continuous improvement of knowledge and CDSS methods. CDSS interventions and clinical knowledge undergo continuous improvement based on feedback, experience with the system, and data that are easy to aggregate, assess, and apply.

Further research is needed to identify the best way for organizations to share "alert" knowledge and to edit commercial medication knowledge bases to yield clinically valuable knowledge bases.[18] Also, more research is needed to identify which member of the health-care team (physician, nurse, pharmacist, or other) is the best recipient for any particular alert and whether physicians and pharmacists should see the same drug-related alerts.

15.7 CONCLUSION

This chapter is intended to provide a broad but not in-depth overview of clinical decision support systems. The conclusion of a 2001 Agency of Healthcare Research and Quality report stated that "the widespread implementation of successful systems is feasible and will likely become more so as providers and systems increasingly shift to computerized [health] record systems."[49] With the increasing electronic availability of patient-specific data, sophisticated clinical decision support is not only needed, but also within reach. Pharmacists are in a unique position to take the lead in this area. The American Society of Health-System Pharmacists acknowledges this unique position in its Statement on the Pharmacist's Role in Informatics (2007)[50]:

Pharmacists have the unique knowledge, expertise, and responsibility to assume a significant role in medical informatics. As governments and the health care community develop strategic plans for the widespread adoption of health information technology, pharmacists must use their knowledge of information systems and the medication-use process to improve patient care by ensuring that new technologies lead to safer and more effective medication use.

REFERENCES

1. Bates, D. W., Cullen, D. J., Laird, N., et al. Incidence of adverse drug events and potential adverse drug events. Implications for prevention. ADE Prevention Study Group. *JAMA* 1995. 274:29–34.
2. Kaushal, R., Bates, D. W., Landrigan, C., et al. Medication errors and adverse drug events in pediatric inpatients. *JAMA* 2001. 285:2114–2120.
3. Classen, D. C., Pestotnik, S. L., Evans, R. S., et al. Computerized surveillance of adverse drug events in hospital patients. *JAMA* 1991. 266:2847–2851.
4. Kuperman, G. J., Boyle, D., Jha, A., et al. How promptly are inpatients treated for critical laboratory results? *Journal of the American Medical Informatics Association* 1998. 5:112–119.
5. Osheroff, J. A., Teich, J. M., Middleton, B. F., et al. Clinical decision support in electronic prescribing: Recommendations and an action plan (http://www2.amia.org/inside/initiatives/cds/cdsroadmap.pdf). Accessed March 30, 2009.

6. Shortliffe, E. H., and Barnett, G. O. Medical data: Their acquisition, storage and use. In *Medical informatics: Computer applications in health care,* ed. E. H. Shortliffe and L. E. Perreault, 41–75. New York: Springer, 2001.

7. Sim, I., Gorman, P., Greenes, R. A., et al. Clinical decision support systems for the practice of evidence-based medicine. *Journal of the American Medical Informatics Association* 2001. 8:527–534.

8. Institute of Medicine Committee on Quality of Health Care in America. *Crossing the quality chasm: A new health system for the 21st century.* Washington, D.C.: National Academy Press, 2001.

9. Glasziou, P., and Haynes, B. The paths from research to improved health outcomes. *ACP Journal Club* 2005. 142:A8–A10.

10. Cabana, M. D., Rand, C. S., Powe, N. R., et al. Why don't physicians follow clinical practice guidelines? A framework for improvement. *JAMA* 1999. 282:1458–1465.

11. Pathman, D. E., Konrad, T. R., Freed, G. L., et al. The awareness-to-adherence model of the steps to clinical guideline compliance. The case of pediatric vaccine recommendations. *Medical Care* 1996. 34:873–889.

12. Kohn, L., Corrigan, J., and Donaldson, M. *To err is human: Building a safer health system.* Committee on Quality of Health Care in America, Institute of Medicine. Washington, D.C.: National Academy Press, 2000.

13. Fonkych, K., and Taylor, R. The state and pattern of health information technology adoption (http://www.rand.org/pubs/monographs/2005/RAND_MG409.pdf). Accessed March 25, 2009.

14. Sotiriou, C., and Piccart, M.J. Taking gene-expression profiling to the clinic: When will molecular signatures become relevant to patient care? *Nature Reviews Cancer* 2007. 7:545–553.

15. Osheroff, J. A., Teich, J. M., Middleton, B., et al. A road map for national action on clinical decision support. *Journal of the American Medical Informatics Association* 2007. 14:141–145.

16. Williams, S. C., Schmaltz, S. P., Morton, D. J., et al. Quality of care in U.S. hospitals as reflected by standardized measures, 2002–2004. *New England Journal of Medicine* 2005. 353:255–264.

17. Patient Safety Authority, Pennsylvania. Results of the PA-PSRS Workgroup on Pharmacy Computer System Safety. *PA PSRS Patient Safety Advisory* 2007. 4 (suppl 2):1–6 (http://www.patientsafetyauthority.org/ADVISORIES/AdvisoryLibrary/2007/may31_4(suppl2)/Pages/may31%3B4(suppl2).aspx).

18. Kuperman, G. J., Bobb, A., Payne, T. H., et al. Medication-related clinical decision support in computerized provider order entry systems: A review. *Journal of the American Medical Informatics Association* 2007. 14:29–40.

19. Lesar, T. S. Tenfold medication dose prescribing errors. *Annals of Pharmacotherapy* 2002. 36:1833–1839.

20. Kozer, E., Scolnik, D., Keays, T., et al. Large errors in the dosing of medications for children. *New England Journal of Medicine* 2002. 346:1175–1176.

21. Peterson, J. F., Kuperman, G. J., Shek, C., et al. Guided prescription of psychotropic medications for geriatric inpatients. *Archives of Internal Medicine* 2005. 165:802–807.

22. Teich, J. M., Merchia, P. R., Schmiz, J. L., et al. Effects of computerized physician order entry on prescribing practices. *Archives of Internal Medicine* 2000. 160:2741–2747.

23. Shah, N. R., Seger, A. C., Seger, D. L., et al. Improving acceptance of computerized prescribing alerts in ambulatory care. *Journal of the American Medical Informatics Association* 2006. 13:5–11.

24. Payne, T. H., Nichol, W. P., Hoey, P., et al. Characteristics and override rates of order checks in a practitioner order entry system. *Proceedings of AMIA Symposium* 2002. 602–606.

25. Spina, J. R., Glassman, P. A., Belperio, P., et al. Clinical relevance of automated drug alerts from the perspective of medical providers. *American Journal of Medical Quality* 2005. 20:7–14.

26. Peterson, J. F., and Bates, D. W. Preventable medication errors: Identifying and eliminating serious drug interactions. *Journal of the American Pharmaceutical Association* (Wash). 2001. 41:159–160.

27. Paterno, M. D., Maviglia, S. M., Gorman, P. N., et al. Tiering drug-drug interaction alerts by severity increases compliance rates. *Journal of the American Medical Informatics Association* 2009. 16:40–46.

28. Chertow, G. M., Lee, J., Kuperman, G. J., et al. Guided medication dosing for inpatients with renal insufficiency. *JAMA* 2001. 286:2839–2844.

29. Evans, R. S., Pestotnik, S. L., Classen, D. C., et al. A computer-assisted management program for antibiotics and other anti-infective agents. *New England Journal of Medicine* 1998. 338:232–238.

30. Chen, P., Tanasijevic, M. J., Schoenenberger, R. A., et al. A computer-based intervention for improving the appropriateness of antiepileptic drug level monitoring. *American Journal of Clinical Pathology* 2003. 119:432–438.

31. Overhage, J. M., Tierney, W. M., Zhou, X. H., et al. A randomized trial of "corollary orders" to prevent errors of omission. *Journal of the American Medical Informatics Association* 1997. 4:364–375.

32. Lasser, K. E., Seger, D. L., Yu, D. T., et al. Adherence to black box warnings for prescription medications in outpatients. *Archives of Internal Medicine* 2006. 166:338–344.

33. Chen, Y. F., Avery, A. J., Neil, K. E., et al. Incidence and possible causes of prescribing potentially hazardous/contraindicated drug combinations in general practice. *Drug Safety* 2005. 28:67–80.

34. Streptase (package insert) (http://www.emc.medicines.org.uk/). Accessed March 31, 2009.

35. Audet, A. M., Greenfield, S., and Field, M. Medical practice guidelines: Current activities and future directions. *Annals of Internal Medicine* 1990. 113:709–714.

36. Tierney, W. M., Overhage, J. M., Takesue, B. Y., et al. Computerizing guidelines to improve care and patient outcomes: The example of heart failure. *Journal of the American Medical Informatics Association* 1995. 2:316–322.

37. Stiell, I. G., and Wells, G. A. Methodologic standards for the development of clinical decision rules in emergency medicine. *Annals of Emergency Medicine* 1999. 33:437–447.

38. Morris, A. H. Developing and implementing computerized protocols for standardization of clinical decisions. *Annals of Internal Medicine* 2000. 132:373–383.

39. Raschke, R. A., Gollihare, B., Wunderlich, T. A., et al. A computer alert system to prevent injury from adverse drug events: Development and evaluation in a community teaching hospital. *JAMA* 1998. 280:1317–1320.

40. Wessels-Basten, S. J., Hoeks, A. M., Grouls, R. J., et al. Development strategy and potential impact on medication safety for clinical rules: The lithium case. *British Journal of Clinical Pharmacology* 2007. 63:507–508.

41. Mille, F., Schwartz, C., Brion, F., et al. Analysis of overridden alerts in a drug–drug interaction detection system. *International Journal of Quality Health Care* 2008. 20:400–405.

42. HIMMS Analytics. EMR adoption model 2008 (http://www.himssanalytics.org/hc_providers/emr_adoption.asp). Accessed March 26, 2009.

43. Chaudhry, B., Wang, J., Wu, S., et al. Systematic review: Impact of health information technology on quality, efficiency, and costs of medical care. *Annals of Internal Medicine* 2006. 144:742–752.

44. Bates, D. W., Kuperman, G. J., Wang, S., et al. Ten commandments for effective clinical decision support: Making the practice of evidence-based medicine a reality. *Journal of the American Medical Informatics Association* 2003. 10:523–530.

45. Bates, D. W., Cohen, M., Leape, L. L., et al. Reducing the frequency of errors in medicine using information technology. *Journal of the American Medical Informatics Association* 2001. 8:299–308.

46. Ash, J. S., Berg, M., and Coiera, E. Some unintended consequences of information technology in health care: The nature of patient care information system-related errors. *Journal of the American Medical Informatics Association* 2004. 11:104–112.

47. Kawamoto, K., Houlihan, C. A., Balas, E. A., et al. Improving clinical practice using clinical decision support systems: A systematic review of trials to identify features critical to success. *British Medical Journal* 2005. 330:765.

48. Teich, J. M., Osheroff, J. A., Pifer, E. A., et al. Clinical decision support in electronic prescribing: Recommendations and an action plan. *Journal of the American Medical Informatics Association* 2005. 12 (4): 403–409.

49. Shojania, K. G., Duncan, B., McDonald, K., and Wachter, R. Making health care safer: A critical analysis of patient safety practices. Agency for Healthcare Research and Quality (www.ahcpr.gov/clinic/ptsafety). Accessed August 9, 2005.

50. ASHP statement on the pharmacist's role in informatics. *American Journal of Health-System Pharmacy* 2007. 64:200–203.

51. Blumenthal, D., and Glaser, J. P. Information technology comes to medicine. *New England Journal of Medicine* 2007. 356:2527–2534.

52. Garg, A. X., Adhikari, N. K., McDonald, H., et al. Effects of computerized clinical decision support systems on practitioner performance and patient outcomes: A systematic review. *JAMA* 2005. 293:1223–1238.

CHAPTER 16

Data Mining for Pharmacy Outcomes

Robert H. Schoenhaus

CONTENTS

16.1 INTRODUCTION

The introduction of medical informatics and electronic medical records into daily practice has made data mining a useful new tool to support outcomes research. Although clinical documentation is routine for pharmacists, the ability to gather and interpret outcomes data rapidly from different computer systems can be extremely challenging. Nevertheless, such information is often essential for improving patient care, formulary management, and general research. For the purposes of this discussion, "data mining" includes investigative queries of patient records and more complex integration of the larger body of integrated patient data from multiple systems.

The role of the data miner has often been delegated to members of hospital decision support teams that may have little or no clinical expertise. This poses the typical challenges of miscommunication between what the clinician may want and what the data miner delivers. Ideally, a clinician with the expertise necessary to harvest information from the existing databases can provide a valuable service to his or her department and to other departments collaborating on outcomes projects.

Data mining can be a complicated process, but some basic skills can make pharmacists less dependent on others. This includes familiarity with how data are structured in clinical documentation systems, how to ask the right questions to get what is needed from a database, and how to structure the data so that they tell the story that is needed. Along with these skills, some rudimentary knowledge of database queries, spreadsheet manipulation, and computer graphics is needed.

The goal of this chapter is to provide some basic understanding of the uses of data mining within a healthcare environment. Such information will guide the development of appropriate outcomes projects to help deliver meaningful information to healthcare decision makers. The expanding role that electronic documentation plays in the daily work of the clinical pharmacist, both now and in the future, justifies the importance of pharmacists as data miners.

16.2 WHERE DOES THE INFORMATION "LIVE"?

Once pharmacists realize that their productivity and research may depend on their ability to gather relevant data, they are typically eager to acquire information. It may seem like a simple task because clinicians expect that everything they see going into a system is easily retrieved from the system in an orderly fashion. Rarely, though, is this the case. Documentation in the health record, whether it is demographic, clinical, or financial, is often routed to separate systems that may not interface with each other in a logical, seamless fashion. This results mainly from different medical departments having different priorities for what they want to do with the data. For example, a financial decision support employee may only want to see billing codes relevant to a patient encounter rather than all the clinical documentation that exists in progress notes.

Historically, storing data in separate medical center systems was of little consequence. However, with the growth of integrated managed care, the necessity to provide information to support reimbursement that is based on performance is greater than ever. In fact, many payers are moving toward reimbursement structures that require certain criteria to be documented prior to issuing any payment. Along with increasing regulatory pressure to provide proof of clinical intervention, access to outcomes information is essential to hospital financial success.

Data mining begins with a basic understanding of databases (e.g., Microsoft Access, Paradox, Oracle) and how to query those databases and subsequently pass the query results to additional programs that filter, sort, and graph the data for further analysis. Training in the use of such programs is typically not part of pharmacy curricula, but outside courses are often available. The extra training may seem like a burden initially, but it will reap great rewards later. An understanding of the data mining process will allow the pharmacist to formulate research questions that are both feasible and rewarding. Being able to answer these questions quickly and efficiently can make a pharmacist very valuable to healthcare administrators.

Where does one begin when looking for data to analyze a research question using a history of patient encounters? What was once in a fragmented paper chart of patient information now likely resides in various databases within the institution. To better understand how data get into these databases, we begin by exploring how data are captured in a typical patient encounter.

16.3 THE PATIENT DATA JOURNEY

From the moment when a patient enters the hospital, the trail of patient data begins. In fact, the data record for patients may begin well before the patient arrives. This may occur with trauma patients en route or pretransplant patients who have just gained access to an organ, where previous clinician documentation may influence the new care plan. Initially, information gathered is likely to be demographic and related to the admission diagnosis. This will normally include, but is not limited to, admitting diagnosis, patient status, vital signs, and payer status. The latter reference may be surprising, but it is quite important to know what insurance, if any, the patient has because it may have an impact on the treatment plan.

Pharmacists may not appreciate the need for such initial admission data, but it can become quite useful in the context of quality improvement and outcomes research. For example, it may be important to know the percentage of patients admitted through the emergency department that are without any insurance coverage. Such patients may present a large financial burden to the hospital that must be compensated for in other areas of hospital practice. Better understanding patient flow through the urgent care process may facilitate a change in practice that allows more patients to be seen and routed to the appropriate practitioners for follow-up.

Understanding the basic demographics of the admission pool is also very useful when planning future care plans. For example, the discovery of a trend of community-acquired methicillin-resistant *Staphylococcus aureus* infection admissions may trigger the infection control service to implement stronger contact precautions for hospital staff and to direct interventions to patients with certain demographics.

It would be rare for patients to make it very far into the healthcare admission process without receiving medications. At this point, pharmacists should be fully engaged in the patient encounter. This includes validating appropriate medication orders, adjusting dosages, and documenting interventions in the care plan. All pharmacists would see the benefit of being able to retrieve data for clinical purposes readily. Just as any pharmacist in any setting would evaluate a medication profile for inappropriate combinations of drugs, so might other healthcare workers or administrators who have access to the data. Communication within an electronic health record is an essential part of modern healthcare. Physicians, nurses, dieticians, pharmacists, and others are all responsible for documenting their various interactions with the patient.

Once pharmacists have entered information into the patient's electronic record, their work cannot be separated from the larger pool of information. Every order that a pharmacist validates typically carries documentation of who signed off on it and any notes that may have been made about the drug. Other professionals practicing in the institution may have access to some or all of these data, depending on the system's architecture and permissions.

Each time a warning message appears about a potential drug interaction or other drug-related problem, the pharmacist who accepts the order and allows administration of the medication is assuming risk and responsibility. Such acceptance is normal in clinical practice, of course, but the value of being able to harvest documentation of appropriate care later becomes more important. Pharmacists are immune neither to the increasing regulatory requirement to provide proof of adequate clinical performance nor to the potential liability of malpractice, which may include failure to document care adequately.

A single patient admitted to the hospital can generate an enormous amount of data, especially if he or she has an extended length of stay. Every drug ordered, every laboratory test done, and every procedure performed are captured somewhere in the volume of patient encounter data. Even how many times the patient goes to the bathroom may be an important part of the health record! Obviously, the priority of collecting all these data is to ensure safe and appropriate care, but many pharmacists do not realize the driver for much of this data gathering is related to financial management. It is not uncommon for hospital finance departments to itemize charges for everything during a patient stay, including how many doses of medication the patient took, how many minutes were spent in the operating room, etc. Although this may seem like intensive "bean counting," it results in a wealth of information that can later be mined for important financial and clinical outcomes.

Only upon discharge or death does the creation of patient data cease. At that point, the patient encounter is closed and becomes part of the hospital database. Commonly, the encounter-specific data are also carved up and sent in different directions for billing and regulatory compliance. Codes related to every diagnosis, procedure, and medication given are filed into separate systems and distributed to hospital administrators in the form of quality assurance reporting. One can only hope that all the information is properly coded to ensure that accurate pictures of the patient experience are painted. Data are mined continuously to look for trends of failure and improvement in the hope of providing better care for future patients.

16.4 PUTTING THE PIECES OF THE PUZZLE TOGETHER

The value of having ready access to patient information during a hospital stay is obvious, but the majority of data mining is done after the patient has been discharged. Only at the point at which all the notes have been dictated and all the diagnostic codes have been recorded is there an opportunity to investigate and compare patient care outcomes. The historical analysis of an electronic health record can be an enormously beneficial exercise, although it does suffer from the typical limitations of retrospective study design: It is highly dependent on complete documentation at the point of care and does not allow for factors like randomization to remove investigator biases. For example, investigators retrospectively looking at the influence of a medication on blood pressure may be hard pressed to weed out confounding comorbidities that could also influence this outcome (e.g., trauma, heart disease, renal disease).

Many reasons exist to harvest patient data. The challenge comes when clinicians requesting information do not realize that simple questions can have very complicated informatics solutions. Even the most elegant electronic health records cannot instantaneously answer clinical questions as simple as "tell me whether this patient has ever used this medication in the past." Such a question assumes a continuity of data between various systems, which is rarely the case. In fact, it is rare for different hospitals and pharmacies to share data readily from their separate information systems. In some cases, even different systems in the same hospital may not directly communicate with each other. The lack of data sharing makes it almost impossible to know a patient's complete medical history. The key then is for the clinical informatics specialist to suggest the correct question and generate appropriate queries to extract information from a variety of databases to address another clinician's concern.

16.5 CASE EXAMPLES

To better illustrate the value of applied clinical data mining, it is helpful to use some real-life examples.

Example 16.1: Daily Monitoring for Inpatient Pharmaceutical Care

Practicing evidence-based medicine and making data-driven decisions can be especially challenging for frontline clinical pharmacists. There is a tremendous volume of emerging evidence on both new and old drugs, all of which may drive development of new guidelines and regulatory performance measures. Monitoring select patient populations using existing drug administration records within a hospital environment can greatly assist the pharmacy staff target those who require more intensive medication management. Such reports can also help facilitate projects to improve documentation necessary for reimbursement.

Unfortunately, it is often difficult to identify patients with a chronic disease state like congestive heart failure (CHF) within the hospital. For example, a person with an admitting diagnosis of migraine may also have CHF. If the pharmacist seeks to discover all current hospital patients who have CHF, then running a query on the admitting diagnosis would exclude some patients from the list.

In this example, only the pharmacy order validation database needs to be mined for data related to patients in the hospital. Although many patients with common diagnoses like hypertension, diabetes, and congestive heart failure are typically on the same types of medication for their disease state (e.g., angiotensin converting enzyme inhibitors), the types of medications used can serve as an efficient means of identifying patients with CHF who are currently admitted.

New reimbursement criteria for inpatients with CHF are forcing resource management staff to audit medical records to ensure that documentation required for payment is present. Working with pharmacy informatics, a report can be generated based on active orders for medications that CHF patients typically use (e.g., digoxin, carvedilol). Although the report may capture some patients with alternative diagnoses like atrial fibrillation, it should include all CHF patients. When run daily, this report typically identifies a number of inpatients for the resource management staff to investigate to ensure that the required documentation for reimbursement is present.

Unfortunately, it is common that all the fields that one would like to include in the report are not found in a single table. The various tables from different hospital databases include unique fields not contained in other databases. Therefore, various databases need to be linked to generate the desired report. The tables are connected by fields they have in common to extract all relevant patient information (see Section 4.4.1). Desired report information includes the active CHF medications the patient is using (discontinued orders are filtered out), the location of the patient in the hospital, and the admitting diagnosis. Also included within this query is a designation of the patient's insurance payer. This helps better focus the resource management staff because they can address the payers who require additional criteria to be documented. Figure 16.1 demonstrates how a query for CHF patients might appear in the relational database, Microsoft Access.

Within Figure 16.1 are four specific tables that address various patient data points that get linked together ultimately to generate a meaningful report. The table "SQLUSER_BI1" is designed to capture basic demographic information related to admitted patients (the "BI" refers to bed index). Here we have chosen to capture demographics like the patient's name, medical record

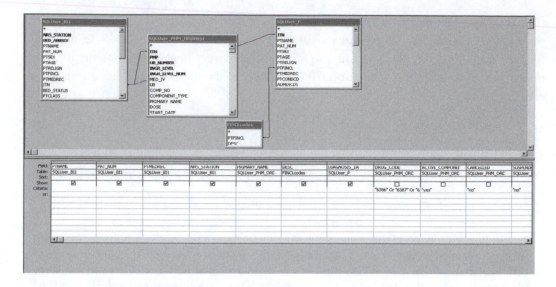

FIGURE 16.1   Query for active CHF patients.

number, bed location within the hospital, etc. The table "SQL_PHM_ORDERS1" is a comprehensive table looking at all medication order information captured. We use this table to identify active orders for the drugs for which we are looking. This is done by selecting drug codes specific to those medications (the codes are demonstrated in quotes in the "criteria" row). The "FINCLcodes" table is added purely to identify each patient's insurance payer.

Note that the SQLUSER_BI1, SQL_PHM_ORDERS1, and SQLUSER_P tables are linked by the ITN number that they all share in common and the FINCLcodes table is linked to the others via the PTFINCL field. The "SQLUSER_P" table contains no specific data that will be printed in the final report, but rather was used to link the financial payer code to the SQLUSER_BI1 and SQL_PHM_ORDERS1 tables.

Example 16.2: Active Warfarin Patients with Laboratory Values

Hospital quality improvement projects commonly target anticoagulation because of the intensive monitoring and potential safety concerns. One way in which pharmacists improve patient safety is to ensure that patients maintain an international normalized ratio (INR) in the target range. The INR should usually fall in the range of two to three. When a patient's INRs fall out of range, a dosage adjustment may be made by the pharmacist or physician to correct the problem. Another example of a daily report that can be used to facilitate pharmacists' clinical activity is a report listing all patients currently on warfarin that also includes their most recent laboratory values. Data mining provides a targeted report for decentralized clinical pharmacists to use daily to make sure that their patient's warfarin dosages are optimal.

Figure 16.2 demonstrates how the query is constructed. In addition to capturing all patients with active orders for warfarin within the hospital, it retrieves the most recent INR values. This helps pharmacists target patients who may be in urgent need of dosage adjustment. The daily report should be distributed electronically to pharmacists via a secure Web site environment. After reviewing the report, a pharmacist can be dispatched from either a central or decentralized location to make a dosage change or counsel the patient prior to discharge. During the patient encounter, pharmacists help educate the patient about his or her medication and dietary habits that may influence the INR. An important patient safety goal is to make sure this happens with every warfarin patient in the hospital.

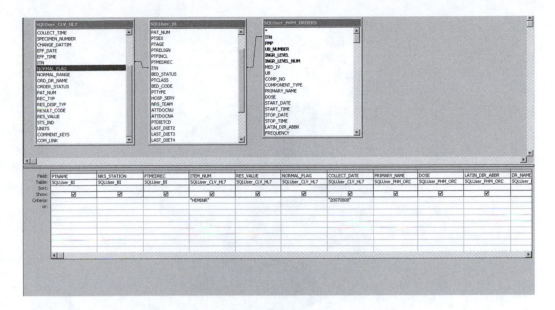

FIGURE 16.2   Query for active warfarin patients with laboratory values.

Figure 16.2 has three tables that are linked to generate a meaningful report. The first table, "SQLUSER_CLV_HL7," includes specific information from laboratories that was ordered for active patients. Here, the table categories related to INR are selected, including the collection date and whether the value was considered "normal." This table is linked to the patient demographics listed in the bed index table, "SQLUSer_BI." The final table, "SQLUSE_PHM_ORDERS," captures active medication orders and is used to identify patients in the hospital on warfarin.

The following more complicated examples detail how data mining can be used to drive quality improvement projects like medication use evaluations (MUEs). These types of projects may be performed by pharmacy staff or pharmacy residents and students as part of their education requirements. MUEs can encompass almost any topic, but commonly target high-risk, high-volume, or high-cost medications used in the hospital. The goal is to assess medication use for appropriateness and determine whether any change in practice is necessary. If needed, changes are often associated with a new guideline or treatment algorithm for staff to follow.

Although not all MUEs tackle complex subject matter, they certainly have that potential if access to data is provided. In fact, the design of MUEs is often aided by data mining for unusual trends. For example, if, upon looking at the hospital's most costly medication, a data miner sees a steep upward trend in the use of a single drug or class of drugs, it may be a legitimate target for an MUE project. The increasing use may be entirely appropriate, of course, but this may not be completely understood until the MUE is completed.

Example 16.3: The Value of Prophylactic Factor VIIa Use

The emergence of recombinant blood clotting products into medical practice has brought great benefit to patients. The products do not carry the risks associated with bovine or human-derived

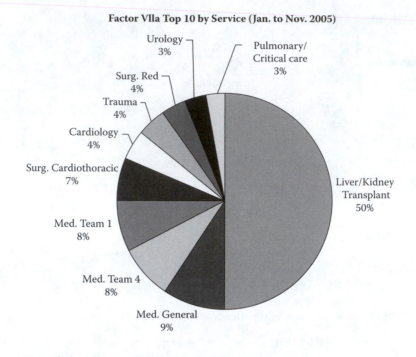

Factor VIIa Top 10 by Service (Jan. to Nov. 2005)

Urology 3%
Surg. Red 4%
Trauma 4%
Cardiology 4%
Surg. Cardiothoracic 7%
Med. Team 1 8%
Med. Team 4 8%
Med. General 9%
Pulmonary/Critical care 3%
Liver/Kidney Transplant 50%

FIGURE 16.3   Use of recombinant factor VIIa by hospital service.

blood products (i.e., infections) and have the capacity to induce hemostasis rapidly in bleeding patients. Trauma and transplant patients, in particular, have been beneficiaries of recombinant factor products. With great benefit, however, often comes great expense. Recombinant factor VIIa is a highly expensive product that was originally indicated only for use in patients with hemophilia. Off-label use for new indications has rapidly grown, causing hospitals to implement guidelines on how it should be used.[1]

Within UCSD Medical Center, the explosion of recombinant factor VIIa use in 2005 and 2006 led to its being one of the top two most costly products used.[2] Although some of this use was related to replacement therapy in patients with hemophilia, the majority was related to off-label indications. To better understand what was driving this process, data on factor VIIa use for the previous year were extracted from databases. These data were easily retrievable from pharmacy databases because they were drug specific. However, the major benefit of gathering such data is to separate them further into use by service or by physician. In this case, the separation of data led to the realization that the liver and kidney transplant service was responsible for the majority of use (see Figure 16.3).

Such data mining examples reflect how easy it can be to direct a quality assurance project. Given the information available, it was impossible to know anything about appropriateness, but investigators did get a logical place to begin asking questions. In this case, use of factor VIIa was specifically related to prophylactic administration prior to transplant surgery. This off-label use had some basis in evidence because previous internal investigations had concluded that an overall reduction in blood transfusion and operating room time might occur when factor VIIa was given prophylactically. Unfortunately, this small internal study had many limitations.

The emergence of large trials related to the unsuccessful implementation of this practice in other hospitals[3–5] led to a full reevaluation of the practice of prophylactic factor VIIa use by abdominal transplant services. This reevaluation was also partially driven by the growing concern of thromboembolic risk associated with factor VIIa; this concern resulted in additional warning on the product labeling.[6] In addition, case reports were beginning to emerge about associated embolic events in liver transplant patients exposed to factor VIIa. Although thrombosis is a known risk of abdominal transplant procedures, it is nearly impossible to determine that a surgical thrombosis was not related to the use of the prothrombotic factor VIIa when given prior to surgery. These safety concerns, along with the lack of supportive data for decreases in blood use or operating room time reduction, resulted in the pharmacy and therapeutics committee recommending discontinuation of prophylactic factor VIIa use in abdominal transplant procedures.

The decision to change medication practice, whether for efficacy, safety, or financial reasons, should always be followed by an outcomes project to verify that the decision was appropriate. In the case of restricted use of factor VIIa by the abdominal transplant service, a data-driven analysis of patient outcomes was initiated to validate the new practice. This analysis evaluated historic data on patients who had received factor VIIa compared to those who did not, controlling for the varying severity of illness of patients undergoing transplant. This retrospective study revealed no appreciable differences in successful transplantation of patients who did not receive factor VIIa. The remaining questions centered around the ability of prophylactic factor VIIa to "pay for itself" in terms of overall cost of patient care because it supposedly reduced operating room time and the need for blood transfusion. Such outcomes might be difficult to demonstrate without access to data typically gathered only by the financial department.

The total cost of care of any patient encounter is typically composed of various contributory components. These include the cost of accommodations (i.e., length of stay), laboratory tests, radiology examinations, pharmacy costs, etc. These components are commonly referred to as "cost centers" of the hospital. Fortunately, for the purposes of this outcomes study, these cost centers also include those for the blood bank and the operating room. The blood bank cost center captures all units of blood products administered per encounter, including fresh frozen plasma, platelets, etc. The operating room cost center captures the total time billed for each procedure by minute. These important outcomes for demonstrating the impact of reducing factor VIIa use in abdominal transplant ultimately helped validate that no financial savings were associated with prophylactic factor VIIa.

The next example represents an even broader approach to answer local safety questions. If possible, it should always be a goal to publish outcomes generated to ensure appropriate medication use at one's hospital. Chances are that other institutions may also be struggling with the same issue. Although it involves more work, publication preserves one's work in the pool of public scientific information for an extended period of time, giving the biggest return on the investment.

Example 16.4: Exploring the Safety of Antifibrinolytics

Sometimes, the ability to detect the safe, efficacious use of a medication appropriately cannot be accomplished with single-center patient data alone. This is especially true when the medications in question are only rarely associated with adverse events. Unfortunately, this is not an uncommon occurrence. Many popular drugs have been withdrawn from the U.S. marketplace only after exposure in millions of people has revealed serious safety problems that were not readily apparent in smaller efficacy trials. Such was the recent case of the medication aprotinin, an antifibrinolytic used to prevent bleeding in cardiothoracic surgery.

Available since 1993, aprotinin was commonly considered a safe and effective agent for the prevention of bleeding in a majority of patients. It had largely replaced common use of older antifibrinolytics for many surgical centers. Although much more expensive than other agents, it was felt to be superior by many cardiothoracic surgeons. Unfortunately, several clinical trials began to find an association between the use of aprotinin and an increased risk of adverse renal, cardiovascular, and cerebrovascular events.[7,8] Clinicians were reluctant to believe there could be serious safety issues with a drug that had been in use for so long, but the available evidence resulted in some restriction being placed on its use at many medical centers.

The safety of aprotinin was called into question only when a large number of patients were compared; thus, it made little sense to assess its appropriate use within a single medical center. Instead, data from 20 academic medical centers were pooled using an administrative database of academic health centers in the University Health Consortium's (UHC) Clinical Resource Manager (CRM) Database. The CRM gathers data from participating hospitals' inpatient discharge summaries, uniform billing (UB-92) forms, charge master records, and detailed billing files.

A search was performed of all hospitalized patients within the medical centers who received either aprotinin or the alternative antifibrinolytics: aminocaproic acid or tranexamic acid. Because this involved analyzing data from patients admitted to other hospitals, patient identifiers were masked to maintain privacy. Patients who received an antifibrinolytic agent and underwent cardiothoracic surgery as defined by their diagnostic related group (DRG) were identified; this is a common way to categorize patient encounters following discharge. During the time period investigated, there was large variability in the choice of antifibrinolytic agent at various cardiothoracic surgery centers. Many centers did not report use of any antifibrinolytic agent at all for surgery patients, calling into question the necessity for consistently using an agent like aprotinin if it is truly associated with dangerous adverse events.

Over 60,000 patients were evaluated as part of this study, which demonstrated an increased risk for adverse events, including acute renal failure, inpatient hemodialysis, and death. No significant difference was demonstrated in the ability of aprotinin or aminocaproic acid to reduce blood transfusions in the general cardiothoracic surgery population, although the coronary artery bypass graft subpopulation did seem to receive greater benefit from aprotinin.[9]

The results of the previous study were shared with local pharmacy and surgery staff. Aprotinin guidelines restricting its use to high-risk populations most likely to benefit from its use were endorsed and its use decreased. Alternative regimens, including those for aminocaproic acid, were also included in the guidelines to help encourage their use, if needed. Guideline implementation saved tens of thousands of dollars on aprotinin expenditure at UCSD. Additional savings related to prevention of adverse patient outcomes were likely large, but difficult to calculate precisely. Medication use was greatly improved by the appropriate outcomes data mined from a large administrative database and pooled with similar data from other institutions.

16.6 SUMMARY

Today's patient experience is full of events that generate useful data for a variety of purposes. As more and more hospitals move toward electronic health records, the role of the data miner in healthcare will grow in importance. Clinicians with knowledge of how to capture and manipulate data from institutional databases will help maintain and improve patient outcomes.

The examples in this chapter illustrate how data mining can be used for a variety of purposes, including daily monitoring reports to make pharmaceutical care more efficient, hospital quality improvement projects like MUEs, and data-driven research projects to assess ongoing concerns with medication safety. These projects can improve patient care, generate new medication treatment guidelines, and potentially lead to publication in scientific and professional journals.

REFERENCES

1. O'Connell, K. A., Wood, J. J., Wise, R. P., et al. Thromboembolic adverse events after use of recombinant human coagulation factor VIIa. *JAMA* 2006. 295:293–298.
2. Schoenhaus, R., and Daniels, C. Applied pharmacoeconomics in academic medicine: Validation of concept. *Value Health* 2007. 10:A202–A203.
3. Lodge, P., Jonas, S., Jones, R. M., et al. Efficacy and safety of perioperative doses of recombinant factor VIIa in liver transplantation. *Liver Transplantation* 2005. 11:973–975.
4. De Gasperi, A., Baudo, F., and De Carlis, L. Recombinant FVII in orthotopic liver transplantation (OLT): A preliminary single center experience. *Intensive Care Medicine* 2005. 31:315–316.
5. Planinsic, R., Van der Meer, J., Testa, G., et al. Safety and efficacy of a single bolus administration of recombinant factor VIIa in liver transplantation due to chronic liver disease. *Liver Transplantation* 2005. 8:895–900.
6. FDA Web site: http://www.fda.gov/cber/products/novoseven.htm (accessed February 12, 2008).
7. Mangano, D. T., Tudor, I. C., Dietzel, C., et al. The risk associated with aprotinin in cardiac surgery. *New England Journal of Medicine* 2006. 354:353–365.
8. Karkouti, K., Beattie, W. S., Dattilo, K. M., et al. A propensity score case-control comparison of aprotinin and tranexamic acid in high-transfusion-risk cardiac surgery. *Transfusion* 2006. 46:327–438.
9. Matuszeweski, K., Schoenhaus, R., Bonk, M. E., et al. Use and outcomes of antifibrinolytic therapy in patients undergoing cardiothoracic surgery at 20 academic medical centers in the United States. *Pharmacy & Therapeutics* 2008. 33:98–106.

V

The Future of Pharmacy Informatics

CHAPTER 17

Pharmaceutical Sciences in a Virtual World

Philip E. Bourne

CONTENTS

17.1 INTRODUCTION

The way in which we learn is changing. Some would argue that these changes, wrought in large part by the Internet, are as fundamental as was the introduction of the printing press in the fifteenth century. In terms of traditional scholarly communication as defined by books and scientific papers, the Internet has offered an alternative system for instantly delivering what was previously in private form only. More recently, however, the full power of the medium is being realized. "Wikis," as exemplified by Wikipedia, bring the wisdom of crowds to bear; students and professionals are communicating in new ways, and rich media that include sound and video are becoming mainstream in education.

These fundamental changes affect the education of pharmacists and how practicing pharmacists and pharmaceutical scientists are working and will work on a daily basis. Such changes are part of the motivation for this book and the discussion of such changes can be found in every chapter. This chapter is devoted to a discussion of some of these fundamental changes as they relate to how we learn and communicate and it draws heavily from our own efforts in scholarly communication to provide examples. If, at the end of the chapter, the reader better understands the changes that are here and on the way and is excited by the prospect of embracing and perhaps even contributing to that change, I will consider it a success.

Let us approach our discussion of the future of scholarly communication by way of a scenario:

> At home, prior to leaving for the university, Jane, a pharmaceutical scientist, syncs her IPOL with the latest papers and video feeds delivered overnight via RSS feed. On the bus, she reviews the stream, selecting a video paper on a recent clinical pharmacogenomics study of warfarin. The data show apparent anomalies. By the time the bus stops, she has recomputed the results, discussed them online with colleagues, proven the anomaly, written a rebuttal, and sent it to the open archive for community scrutiny.

Some would say this example is science fiction. I would argue that some aspects of this scenario are already in place and the rest a few years away at most. Let us break it down to see whether I can be convincing that this represents the future.

17.2 THE INFRASTRUCTURE IS IN PLACE

The term *IPOL* is intended to describe an iPod–laptop hybrid that effectively already exists in that an iPod can be used as a laptop and a laptop can be used as an iPod. Handheld devices are an integral part of pharmacy practice and pharmaceutical sciences, as we saw in Chapter 12, and are becoming more sophisticated all the time. With the open interfaces found on the iPhone and other devices, the number of applications will continue to grow in number and sophistication. Thus, as tools, handheld devices will become an increasingly important part of one's professional life.

17.3 THE PUSH–PULL OF INFORMATION

"…delivered overnight via RSS feed." When one types in a URL to a Web browser and a page appears, this is called "pulling" information: The user initiates the download of information from a central server. "Pushing" of information is when the server sends information to the user without their explicitly asking for it. Really simple syndication (RSS) is one example of this phenomenon. It is already used for a variety of purposes for receiving appropriate information on a regular basis without remembering to ask for it.

A typical use is to be informed of new research papers or data as they become available on a central server. RSS feeds are a good way to get the table of contents of a favorite journal delivered as soon as the issue is published. It is not necessary to go to the journal Web site to retrieve it. Thus, these kinds of alerts can be valuable because, when they are viewed in a Web browser that supports an RSS feed or in an RSS reader, they provide awareness of new information as it becomes available. Such constant updates are reminiscent of a ticker that constantly scrolls across the bottom of a television screen providing immediate updates to things that change frequently, such as stock prices.

17.4 RICH MEDIA AND MASHUPS

"…video paper on a recent clinical pharmacogenomics study of warfarin." The idea of a video paper is unusual, but yet it makes perfect sense. Much of what we read can be expressed and comprehended better in a visual format. The only impediment is that we are not used to this and it does not imply the same credit to the author. Academic advancement comes from writing high-quality scientific papers, not from making videos. Yet much of what we comprehend from a paper could be expressed much better when seen in action rather than reading about that concept. Consider this in the context of a scientific experiment. In many instances, it is easier to see an experiment performed than to read about it in the methods section of a paper. The *Journal of Visual Experiments* (www.jove.com) is proof that this can work. Jove is a video journal indexed by PubMed that is an example of effective use of rich media in science.

SciVee* (www.scivee.tv) is yet another experiment in this direction. The motivation for SciVee is to improve comprehension of and interest in science. A reader of a clinical or basic science paper can peruse the abstract in a minute or two. If the paper sounds appealing, it could take upward of 3 hours to fully absorb its content. When 50,000 papers are being added to PubMed each month, even keeping up with the small subset of the literature in which one is interested becomes a daunting task. SciVee has the notion of a "pubcast," which is typically a 5- to 10-minute video clip by one or more of the authors describing the most salient features of the paper. The video clip is integrated with the paper itself, so, as the author talks about a specific aspect of the paper, that part of the paper (typically figures or tables) appears on the screen.

An alternative is to read the paper on the screen and when a confusing part of the work is encountered, click on the paper to see the author pop up and discuss that specific aspect of the work, which will make it clearer. An example of this can be found at www.scivee.tv/pubcast/16244704. SciVee also supports "postercasts." This is a similar idea, but integration is via a poster with the author describing the contents rather than by a scientific paper. Similarly, "CVcasts" integrate a curriculum vitae with a video of the owner describing his experiences and skill set. Pubcasts, postercasts, and CVcasts are examples of "mashups": integration of multiple information streams. Mashups are already common in the Web 2.0 environment,† but have yet to be widely adopted in the process of scholarly communication.

Such mashups bring up the issue of copyright and general accessibility to the online content of a journal article or book. Section 17.6 discusses open access as a major catalyst for changing how a pharmacist will gain access to information in the future.

17.5 DATA AND KNOWLEDGE NEED NOT BE DISTINCT

"The data show apparent anomalies. By the time the bus stops, she has recomputed the results." At this time, much useful clinical and basic science data are available to pharmacists through a variety of online databases. This availability provides much of the impetus for this textbook and for courses in pharmacy informatics worldwide. The issue that prevents our scenario from being a reality today is that the knowledge derived from these data is contained in the literature and is not integrated with the data upon

---

* The author is the cofounder of SciVee.
† Web 2.0 implies the second generation of the World Wide Web, characterized by interoperability, and the use of video, sound, and communication between users.

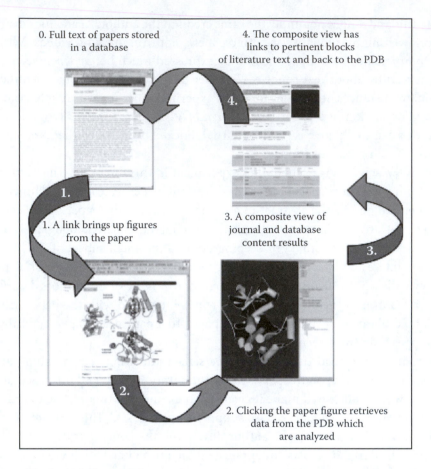

FIGURE 17.1 The knowledge and data cycle.

which it is based. However, we are seeing the first signs of this integration as biological and clinical databases are becoming more like journals and journals are becoming more like databases (see www.ploscompbiol.org/article/info%3Adoi%2F10.1371%2Fjournal.pcbi.0010034).

Let us consider this point a little further. Users are requiring more than basic data from databases. They are expecting annotation of the data. That is, rather than just raw numbers, users expect details of the implications of those numbers. For example, rather than raw statistics on a clinical study, users want details of what those numbers imply for the particular treatment. Conversely, journals are including more and more primary data as supplemental information to the paper. Often these data are unstructured and not easily interpreted by computer. These journals are often online and, depending on their policy (see Section 17.6), the full text may be freely available to all.

Consider a future when data and the literature are fully integrated. An example different from our warfarin scenario illustrates the full power of the approach for data that can be represented in a particularly pictorial way. In Figure 17.1, each arrow represents a step in scholarly learning in a world where data and the knowledge derived from these data are integrated:

Step 1 is familiar now; in this case, the relationship between the structure of a biological macromolecule and its activity is being studied. Within the Web page of the journal article, the reader can click and see a picture of the molecule—in this case, a protein kinase—and attempt to better understand its function. However, such a static, two-dimensional view is not a particularly useful way in which to study structure–activity relationships.

In step 2, the reader clicks on the figure and a new window appears with a renderable view of the molecule as found in a variety of molecular graphics programs. Consider how that figure was made to appear. The click initiated a Web service call (a way for applications to communicate on the Web), whereupon the three-dimensional Cartesian coordinates were retrieved from a database. Metadata (data about the figure) deposited by the author of the paper, but not viewed as part of the article, defines how to render these data in a way that represents the molecule very similarly to how it appears in the journal article. Now, rather than a static two-dimensional image, a "living" three-dimensional object is present from which much can be learned. A seamless interface has been established between a scientific paper and the data in a database upon which that paper is based.

Step 3 illustrates how the figure itself becomes an object of inquiry. It can be probed so that, in the protein kinase scenario, a region of the molecule has a glycine-rich loop. Selecting this feature returns a composite view of information taken from a variety of databases and papers pertaining to this feature.

Step 4 completes the cycle by returning to another paper that may not have been discovered without data–literature integration, and the cycle continues. The end result is another mashup; in this instance, however, rather than literature and video, the reader has access to literature and data.

It is easy to imagine how this concept can be taken further. For example, each figure in a paper can link to an underlying spreadsheet in which readers (i.e., users) of the figure can reprocess the data in ways perhaps not conceived by the original authors. Admittedly, this will likely require a shift in ideals because many scientists protect data and will not release it into the public domain. I would argue that if public money were used to generate these data, they should be freely available to the public. This takes us to the general topic of openness, an important driver of a new virtual world.

17.6 OPEN ACCESS

"…sent it to the open archive for community scrutiny." Science, technology, and medical (STM) publishing is undergoing change as publishers grapple with moving from a predominantly print-based medium to an online-only one. Many scientific journals are now online only and this trend is expected to continue. The potential availability of full text online, the increased cost of journal subscriptions, and the vision of some who saw scenarios similar to the one presented earlier, led a few to propose a new publishing paradigm:

an open-access model, which rather than a "reader pays" model, is an "author pays" model. This does not work for every author because of the cost. However, agencies that pay for the research are now also willing to pay for publishing the results of the research. This still might not work for some research done without grant funding, as can be the case in pharmacy.

Open access was pioneered by organizations like BioMed Central (BMC) and the Public Library of Science (PLoS). Over 4,000 open-access journals now exist, according to the Directory of Open Access Journals (DOAJ; www.doaj.org). BMC was recently bought by Springer Science and Business Media and continues to thrive—a testament that the open-access model can work. The power of what open access brings requires that we examine the license under which open-access materials are made available. These arrangements vary; BMC is published under a creative commons (CC) 2.0 license and PLoS under the most general CC 3.0 license.

Both licenses imply that the copyright remains with the author, rather than being signed over to the publisher. In retaining the copyright, the author agrees that anyone can use the material for any purpose he or she chooses, provided that the original authors are attributed. This includes the mashups illustrated previously. The difference between CC 2.0 and CC 3.0 is that 2.0 implies that this can be done only by nonprofit organizations, whereas 3.0 implies that anyone can do it and try to turn a profit if that is desired. As with any license, there is a lot more to the legal side. Those wishing to read more can refer to creativecommons.org.

Another aspect to open access must be emphasized. Not only is full text made available through resources like PubMed Central (PMC; www.pubmedcentral.nih.gov), but some of it is also made available in extensible markup language (XML), a form of markup that has the potential to enrich the content semantically for use by computers to bring new meaning to that content. A simple example would be to tag references to database content within the body of an article so that it could be retrieved in the kind of scenario illustrated in Figure 17.1. In this way, the paper just becomes another view of the data in a database.

Chapter 4 described how informaticists go to a great deal of trouble to develop controlled vocabularies, taxonomies, and ontologies that describe a particular domain. Tagging such terms in the literature improves searching and makes useful associations between items of content. Ideally, authors would perform semantic tagging as they write the paper, but that is asking a lot. Postprocessing of existing text to add semantic content can still be useful and can lead to new discoveries. For example, suppose a discussion of two genes appears in the same paper and in a number of other papers spread across the scientific literature. Any single reader might never see the relevance of the association, but a computer capable of parsing and looking for associations in millions of papers would find that association, provided the two terms could be easily identified within each paper. Such is the potential for the use of semantic enrichment. Reviewing how papers from PMC are analyzed for such associations can be found in resources like BioLit (biolit.ucsd.edu).

Currently, much ignorance remains around open access, in part because too few applications exploit the content. Most authors of content in open-access journals have not contemplated that someone else can take their intellectual property and effectively repackage it. Authors are not likely to allow this (by publishing in closed-access journals) unless they see that it adds quality without having an impact on their original intent. It will be interesting to see how this plays out in the future.

Many scientists also perceive that open access implies lower quality. This is not generally true because there is no change in the peer review. *PLoS* illustrated this in dramatic fashion when *PLoS Biology and Medicine* revealed its first impact factors of over 10*—a level that most closed-access journals never reach.

A driver for open access is data showing that, on average, an open-access article will be downloaded and cited more than a closed-access article. This widespread dissemination is certainly an incentive for authors because it affects their h-index†—a (somewhat controversial) number that purports to define scholarly success. Many funding agencies around the world that support scientific research are mandating open access for the science they support, typically after 6–12 months of closed access. Needless to say, traditional closed-access STM publishers are fighting these initiatives because they will likely cut into profit margins.

The more forward-looking publishers and information scientists are experimenting with new concepts made possible with the availability of unbridled full text. Let us consider a couple of these concepts. *PLoS* has introduced the notion of collections and Elsevier the idea of an information hub. Both serve to redefine the notion of a traditional journal to that of a more general source of information on a given topic. Collections tie together content from multiple journal sources and provide the community a means to interact online around that content. Hubs bring data, dialog, and journal content together around a common theme. Dialog is a key element here and takes us to the last part of our scenario.

17.7 PROFESSIONAL NETWORKING

"…discussed it online with colleagues, proven the anomaly, and written a rebuttal." Perhaps the greatest phenomenon surrounding the Internet is the ability to communicate in new ways. When I poll first-year doctor of pharmacy students as to who uses Facebook, the number has now risen to almost everyone (57 of 60 in a recent class). A daunting situation one year was that most students were on Facebook and none of the faculty, except for me, were: The digital divide in full force. Now a small group of the younger faculty uses Facebook regularly. Interestingly, when I poll students to determine their interest in using social networking tools like Facebook for professional use, many shy away. Perhaps not surprisingly, they wish to keep their personal and professional lives separate. Digging deeper, it turns out that they are not averse to using online networking tools for professional purposes, but not those that merge social and professional activities.

Indeed, separate resources have emerged to support professional networking. LinkedIn (www.linkedin.com) is a popular resource among science professionals. Other sites are starting to emerge that go beyond linking people together to enable them to exchange

---

* An impact factor of 10 implies that, on average, each paper in a journal is cited 10 times. For further details on impact factors for journals, see www.thomsonreuters.com/products_services/scientific/Journal_Citation_Reports#tech_specs.

† An h-index of 20 implies that an author has 20 papers with more than 20 citations. For further details on h-index, see en.wikipedia.org/wik.

information around specific subject areas, store data, express ideas, and so on. It is too early to tell whether this represents a new way to do science.

A shift toward more openness does seem to be occurring, which is a hallmark of the Web 2.0 generation. Scientists once protected their data, squeezing every last piece of knowledge from it before making it publicly available, if they made it available at all. Now, with scientists undertaking more collaborative projects and with funding agencies requiring that data be made available, some changes are occurring. In the extreme, the complete scientific discourse could be available online and available for colleagues to see and respond to. As soon as a postulate is made, it will be posted—as will the protocol used to test that postulate and then the results. Finally, the conclusions associated with the basic science or clinical study (protecting patient privacy, of course) will be published.

In short, the emerging paper will be available as the work is being done and the paper is being written. One could even imagine a day when the contract between publisher and author is different. Rather than accepting the final finished product, the publisher would be the central repository for all aspects of the study—an archive as well as a disseminator. More science fiction? Perhaps, but take a look at Nature Proceedings (http://precedings.nature. com/), a repository for users. It is already used to store presentations, posters, and even as a registry for work that will be published, which reflects when it was done and by whom.

One manifestation of this openness and collaborative spirit is shown in the idea of a wiki, where multiple contributors come together online to provide a collective wisdom, a so-called "wisdom of crowds" approach to disseminating knowledge. Wikipedia, used by many of us every day, is a startling example of the success of this approach. The collation of information on any given drug is quite remarkable—just one aspect of a resource that is deemed as accurate as the *Encyclopedia Britannica* (www.nature.com/nature/journal/ v438/n7070/full/438900a.html), but with much greater scope and depth.

The idea of using collective wisdom is not new. In *The Professor and the Madman* (see Chapter 4 and www.amazon.com/Professor-Madman-Insanity-English-Dictionary/ dp/006099486X), one can learn how a large body of people contributed to the *Oxford English Dictionary*. Rather than paper notes received by mail and collated and checked by a small group of annotators and curators, we have a wiki with hundreds of thousands of contributors of content who also act as curators and annotators. I would argue that an impediment to these kinds of contributions remains: lack of an appropriate reward structure. Contributions to wikis, blogs, and databases rarely are rewarded when measuring the scholarly output of an individual. This may change when those comfortable with the digital medium are the ones rewarding the next generation of upcoming scholars (see www.ploscompbiol.org/article/info%3Adoi%2F10.1371%2Fjournal.pcbi.1000247).

Already I have seen a few blog postings on a scientific paper that are better than the formal reviews that were written. Perhaps these will remain isolated cases, or perhaps it speaks to a breakdown in the traditional peer-review process. Time will tell and change will likely be different for different communities. For example, physicists are comfortable depositing their scholarly work in the arXiv (arxiv.org), where the peer review is essentially a function of the number of people who download, read, and comment on the work in an open forum. Firmly engaged in the editorial and review process as carried out by traditional journals, biologists are not ready for this yet.

17.8 SUMMARY

Using a scenario of how we might learn and interact in the future, I have introduced a few of the approaches to new modes of scholarly communication that have the potential to have an impact on how pharmacists and pharmaceutical scientists learn and, indeed, disseminate information. I doubt that anyone could have predicted 10 years ago that Facebook would emerge as a communication phenomenon with 150 million users worldwide, or that Wikipedia has the potential to be an online repository for the world's collective knowledge. Perhaps all that is sure is that equally remarkable uses of information technology will emerge in the next 10 years. Are you ready?

CHAPTER 18

Where Do We Go from Here?

Richard M. Peters

CONTENTS

18.1 INTRODUCTION

The fundamental question for all of us as clinicians involved in the continuum of patient care is, What the ultimate goal of information technology and automation in healthcare? In many industries, the use of information technology has very concrete value, often focused solely on increasing efficiency and operational management, with the ultimate goal of increased productivity. In healthcare, however, we are conditioned to look somewhat askance at efficiency for productivity's sake. The reason is that, when we deal clinically with the quirky, simplicity-defying, organic quandary of the human body, we are all too often stymied in our efforts to be efficient. In fact, our clinical experience often shows us that the disease process has its own timeline far outside our "9 to 5" expectations.

That said, efficiency and productivity are critical issues in many aspects of the clinical care process, particularly in the realm of pharmacologic therapeutics. The benefits of early intervention (such as early administration of antibiotics in community-acquired pneumonia)—even outside the critical care setting—demand efficiency in inpatient dispensing.[1] Equally important from a business perspective, automation of dispensing and work-flow productivity are both critical issues in the outpatient pharmacy setting. The information technology tools to achieve efficiency and accuracy in these settings have been discussed in previous chapters. However, healthcare requires a broader definition of information technology and goals that might be less concrete, but no less important.

18.2 GOALS

One of the primary early clinical drivers of healthcare information technology and electronic health records (EHRs) was the provision of automated decision support. The concept was that human clinical decision making could be replicated and improved by computer logic, increasing diagnostic accuracy and improving clinical outcomes. This elusive goal is still a driver of health information technology 40 years later, but it must be put into what is now a broader set of goals that we can clearly articulate:

Decision support. This goal is elusive, but still a primary driver, although our focus has shifted from machines thinking for humans to the use of information technology to provided assisted decision support at the point of care.

Improving patient care quality and accuracy. One of the overarching assumptions in healthcare information technology is that better and more complete information at the point of care, with or without additional decision support, will improve quality and outcomes.

Increasing efficiency and lowering costs. As for other industries, the assumption in healthcare is that automating the work-flow process and information management will provide efficiencies and increase productivity.

Improvements in overall public health. Aggregation of data across populations should allow us to better understand epidemiologic trends and isolated and triggering events, as well as provide overall improvements in the health of the public at large.

Research. As with public health, health information technology should facilitate the aggregation of data for retrospective analysis and the mining of population data, as well as structured management of clinical trials for prospective outcomes research.

Disease management. It is hoped that using information technology in healthcare to involve the patient (and the family when appropriate) to a greater extent in his or her care and in understanding and intervening in his or her disease or wellness process will lead to improved outcomes, lower incidence of disease, and lower overall healthcare costs.

Standardized care. There is an assumption that information technology can help facilitate evidence-based standardization of care.

Communication. Healthcare requires the use of diverse lines of communication and the communication of complex concepts and information—things that electronic communications should facilitate.

Long-term improvements in efficacy. Operational, diagnostic, therapeutic, and interventional improvements should be possible. The feedback loop that information technology can provide relative to interventions and their outcomes should lead to long-term systemic improvements in healthcare.

Education. Information technology in healthcare should facilitate real-time, point-of-care, retrospective, and prospective education for students in the health professions, professionals, and patients.

Accountability. Healthcare information technology should provide a high degree of accountability to all providers, interventions, and decisions across the continuum of care.

For all of these goals, the devil is in the details and, as discussed in the preceding chapters, the details are complex. In trying to frame where we should go from here in health information technology, one of the most important tasks is to simplify that inherent complexity.

18.3 OBSTACLES TO ACHIEVING THE GOALS

The primary problem we face and the source of much of the complexity in healthcare information technology is our insistence to date on automating existing processes within our traditionally established care centers. Our chapter titles in this book illustrate the problem: We have broken the delivery of healthcare, even in pharmacy, into elemental work-flow processes and have failed to take an organic, overarching view of the problem and defined a comprehensive data-driven rather than process-driven solution.

At the heart of this problem is that, like other industries, we have defined information technology as an automation tool for our business processes. That model would be valid if we were building widgets or providing a one-stop service. In healthcare, however, the widget is a healthy patient and the service is a comprehensive set of services provided across a wide array of locations and over a long period of time. The fundamental problem is that we have not made the patient the center of our entire process and his health the core upon which we build our information technology. In essence, we are automating our practices, hospitals, laboratories, and pharmacies, but we are not automating patient care. We have let traditional concepts of automation color our view.

At the center of this quandary is our definition of an electronic health record as a system rather than as an actual lifetime individual patient health record. Instead of patient records, what we have is a set of data silos, centered on the individual computer systems we have implemented across our sites of care to automate our processes. As discussed in previous chapters, this arrangement often makes it extremely difficult to obtain even something as simple as a current medication list for any given patient.

18.4 SOLUTIONS AND IMPEDIMENTS TO IMPLEMENTATION

The solution is to take a data-centric instead of a process-centric view of a health record. This requires a fundamental change in the way in which software engineers and vendors think about health information systems. In almost any other industry, data are proprietary. Data are the source of value and therefore each different company or organization wants to manage its own data. Most software projects, therefore, concentrate on end-user features and functions with the tendency to build a discrete data model and database to support those features and functions.

We have followed this model in healthcare and the result is that even though we can conceptually think of a medication or medication order as identical across locations and across departments, each departmental system has its own database and data model to support its unique needs. Regrettably, a medication in one database is not the same as that identical medication in another. Data messaging standards (HL7, ASTM, NCPCP, ASC X12) and nomenclature standards (SNOMED, CPT, ICD, NDC, RxNorm) were developed to facilitate data exchange and one-to-one translation between disparate systems. (For definitions of abbreviations, see Chapter 4.)

What is emerging in healthcare informatics is the concept that what we need is a standardized data model and database across the healthcare system rather than discrete, departmental- and process-specific data models and databases. A medication or medication order under such a model would be required to be identical across all systems that write to or access the stored data. Note that this model does not mandate that all patient data must be held in a centralized data repository, but rather only that data in storage always be stored with the same parameters, same structure, and same characteristics so that any other authorized systems can interact with those data.

The other fundamental step is to improve the quality and computability of the data we collect in our disparate systems. The majority of current electronic health records collect data primarily as free text and the majority of both claims and electronic prescribing systems in pharmacy currently treat the directions ("sig") portion of a prescription as free text. With its inherent descriptive flexibility, free text is searchable, but it is not computable for exacting processes such as drug dosing and complex administration error prevention (e.g., preventing the administration of contraindicated medications in patients with glucose-6-phosphate dehydrogenase [G6PD] deficiency).

18.4.1 Data Compatibility and Accessibility

If we go back to the list of goals in pursuing automation through the use of health information technology, the optimization of every one requires computable data. Computable data mean structured data. If we limit our discussion just to pharmacy data, we can illustrate the problem in detail.

Medication nomenclatures, like our departmental and process-based information systems, were designed to manage specific functions. They were not designed for data portability and the code sets we need to use in pharmacy decision making were not designed for compatibility.

A good illustration of the problem and a map to solutions can be illustrated with a patient who has G6PD deficiency and allergies to fosfomycin and quinolones, is taking birth control pills, and has a urinary tract infection, end-stage renal failure with a creatinine clearance less than 50 mL/min, and a prescription for Macrobid 100 mg twice daily for 7 days. To prescribe and dispense a medication to treat this patient safely would require data from an EHR for the diagnoses of G6PD deficiency and renal failure; a current medications list and the allergies to fosfomycin and quinolones; data from a laboratory information system for the latest creatinine clearance or serum creatinine; data from an e-prescribing system to evaluate what the physician prescribed; and data from a drug information system for drug dosing, drug–drug interactions, and drug allergy checking.

G6PD deficiency can be discretely described and coded under ICD-9 as 282.2 (G6PD). A urinary tract infection, on the other hand, can be coded as 595 (cystitis), 595.0 (cystitis, acute), 599.0 (urinary tract infection), or 599.10 (acute pyelonephritis). Fosfomycin, birth control pills of unknown type, and Macrobid come with no code, one of many NDC codes, an RxNorm code, or a proprietary drug information code from one of the drug information vendors. Fosfomycin might be coded as a single drug allergy and quinolones as a class allergy in a proprietary EHR drug class, drug information drug class, or an FDA allergy code. The creatinine clearance or creatinine would come with a CPT-4 code, a LOINC code or codes, or a SNOMED-CT code or codes.

The clinical problem is that this patient should not take Macrobid or a sulfonamide due to G6PD deficiency. She should not be given fosfomycin, ciprofloxacin, moxifloxacin, or levofloxacin due to allergic risk, and amoxicllin-clavulanate is best avoided with oral contraceptives. A therapeutic alternative is cephalexin, which would need to be given every 12 hours instead of every 6 hours due to the patient's impaired renal function. However, cephalexin risks treatment failure because of communitywide urinary tract pathogen resistance.

To further define this as a realistic scenario, the contraceptive was prescribed by the patient's primary care physician using a small-practice EHR, and it was purchased by the patient for cash from a discount pharmacy because it is not covered by the patient's pharmacy insurance benefit plan. The urinary tract infection was diagnosed in a stand-alone urgent care center and the patient is using her pharmacy benefit to purchase the antibiotic at a community pharmacy. The patient's nephrologists practice at an academic medical center that uses a large institutional EHR and the patient's laboratory studies are done by one of the large national laboratories, as mandated by the health plan.

A prescribing physician or clinical or dispensing pharmacist really has no choice but to take a complete history from the patient at each step in the process and rely on whatever books or information systems available to reconstruct this patient's therapeutic profile. Otherwise, the expensive EHR and departmental systems are likely to fail and put the patient at risk for a bad clinical outcome.

The ideal situation would be to have a patient- and data-centric record. In other words, any information system that any clinician or ancillary technician, such as a pharmacy technician, touched relative to this patient's care would interact with and keep up to date the patient's personal comprehensive health record. That record, literally or virtually, would follow the patient and be accessible, with the patient's consent, for any interaction with the healthcare system. In addition, that record would be structured, the key data defining that patient's current and historical health status would be structured, and the real-time decision support needed to provide direct care to that patient would be computable.

The obstacles faced in reaching this goal are both practical and political:

- Healthcare organizations and commercial healthcare information technology vendors have invested considerable money and time in systems that do not fit a data- or patient-centric model. Commercial vendors are understandably very wary of adopting an open data model or shared data infrastructure because it opens the health information systems market to smaller vendors and to open-source (i.e., free) software developers who could undercut the larger vendor's market share. Large healthcare institutions in a competitive healthcare marketplace like the United States face the same demon. Opening up or widely sharing comprehensive patient data also makes them vulnerable to competition.

- The larger commercial systems and the publicly funded healthcare information technology systems from the Veteran's Administration, the Department of Defense, or the Indian Health Service are outdated. The majority of them run on MUMPS (Massachusetts General Hospital Utility Multi-Programming System), a database system developed in 1966 at the Massachusetts General Hospital in Boston. MUMPS was developed in an era when computing power (processors) and computer storage resources (memory) were expensive. MUMPS combined efficient storage with a simple, descriptive, and highly configurable programming language. MUMPS is extremely efficient at individual record storage and retrieval, such as an individual patient's health record. It has found wide use in healthcare systems and in banking and financial services.

  However, MUMPS is extremely *inefficient* in retrieving aggregated data across records. This characteristic is not a problem in banking and financial services, where these data are simple, numeric, and aggregated in external data files. However, in healthcare, the majority of data are not numeric and population data across patients are critical for operational management and clinical care. Although large healthcare systems have stuck with MUMPS, banking and financial services have migrated toward relational database management systems for aggregated data and use MUMPS solely for discrete record storage and retrieval.

  Meanwhile, Google, Amazon, Yahoo!, Sun Microsystems, and other large Internet companies have migrated toward large, highly scalable, indexed data storage and retrieval systems that provide both discrete and aggregated data management. In an era when it is necessary to move discrete and aggregated patient data rapidly across a distributed network like the Internet, much of healthcare continues to run on systems designed decades ago.

- Management of cross-system aggregated data, even on a single patient, is problematic. In the 1980s, it was suggested that every patient should carry a "smart" card. These smart cards, similar to a credit card in shape and size, would hold the patient's entire health record. Then, when a patient was seen, any healthcare information system would be able to read the card and healthcare providers would be able to interact with and keep the patient's portable health record up to date. Germany and France conducted the largest trials of this concept, which were regrettably a failure for very simple reasons: (1) The data on the cards and on the associated systems were not discretely structured and standardized enough to be accurate or interoperable, and (2) patients would forget or misplace their cards or, in the case of trauma, would often be brought to the hospital without them.

- The other option is to keep a given patient's data on a centralized or distributed repository and let disparate systems access the data when a patient presents for care. The problem with this approach involves adequately protecting patient privacy, which is our fourth obstacle to patient-centric data. Laws in all countries belonging to the European Union (EU) strictly protect data and individual privacy and confidentiality. This makes card programs in France and Germany acceptable, as are data repository projects.

  The United States, however, has no inherent legal privacy or confidentiality protections for individual data. Policy, as set by regulations like HIPAA, offers regulatory protection and financial penalties for misuse of healthcare data in the United States; however, unlike in the EU, no overarching restrictions prohibit data use. Lacking EU-like data privacy protection, U.S. privacy advocates have effectively blocked data aggregation and data repository projects. Until the United States adopts EU-style privacy laws that apply at least to healthcare data, opposition to large-scale shared patient data repositories (with the exception of consent-based repositories from companies such as Microsoft [HealthVault] and Google [Google Health]) will be opposed.

- For data to be computable, they have to be structured and use a standardized and constrained terminology or terminologies. The political problem is that standards and terminologies in healthcare exist in silos just as our information systems do. NCPDP is used by pharmacies and pharmacy benefit managers, ASC X12 is used by health plans, and HL7 is used by the large existing commercial vendors and publicly funded systems. Meanwhile, the FDA is doing its own work on allergies and adverse reactions and the federal government is funding alternative drug coding with RxNorm. SNOMED–CT is now international, but it is used in few EHR systems. Meanwhile, Medcin, a privately developed and proprietary coding system, is widely used. ICD-9 and ICD-10 are applicable for billing and disease classification, but they are not discrete or structured enough for clinical medicine. Nursing diagnostic and descriptive code sets are out of sync with physician and pharmacy coding systems. LOINC and CPT coding for procedures are lab centric. They also lack clinical relevance and are even too nondiscrete to be used in clinical order entry.

  Each one of the standards organizations and terminology sources—with the exception of LOINC, the FDA, and RxNorm—is in the business of making money, either through the sale or use of its solutions or through membership. The National Library of Medicine (NLM) has done an admirable job of aggregating all the code sets together, but the NLM metathesaurus, while providing term-to-term and code-to-code matching, is the opposite of what is needed clinically. What is needed for clinical decision support and computability is a limited, specific, standardized, and clinically relevant terminology and code set—ideally, from one source.

18.4.2 Incorporating Finances into Clinical Decision Making

The other missing piece in healthcare information technology is financial data that define the cost and expenses associated with the provision of patient care. Physicians and nurses are essentially blind to these data, while pharmacists, patients, administrators, pharmacy benefit managers, and health plans deal with them every day. The problem is that if we are even to begin to control healthcare costs, pricing transparency will be necessary to make financial considerations a factor in effective clinical decision making.

At issue is that physicians are currently making almost 100% of all initial diagnostic, interventional, and therapeutic decisions in healthcare completely insulated from cost. Outside of issues of eligibility or prior authorization for procedures, the only place physicians occasionally run into cost issues is with prescription drugs. This fact makes prescription and acute care drugs an ideal illustration of how to restructure health information systems to support pricing transparency as in the scenario that follows.

In the outpatient setting, when physicians write a prescription on paper, they often have no idea what the patient's drug benefit is. This means that they might have no concept of the formulary or any idea of the patient's copayment obligation. In the paper world, even routine prescriptions often result in a call from the dispensing pharmacist stating that the requested drug is not on the patient's insurance formulary and will cost the patient $150 out of pocket. What tends to occur is a round of phone tag and missed connections, frustrating the pharmacist, the physician, and, even more importantly, the patient. The result is verbal consent from the physician to change the drug and modify the directions as required. What the patient ends up with is a drug different from the one the physician prescribed and discussed with the patient. The burden then falls on the dispensing pharmacist to start from the beginning and re-educate the patient.

The other scenario is that the patient fills the prescription the first time, but due to cost, fills it only intermittently from that point on. It has been estimated that 50% of prescriptions are never filled, and analysis of claims from pharmacy benefit managers has shown that medications for chronic conditions are filled, on average, only 7–8 months out of any 12 months.[2] Compounding the problem, when the patient next sees his physician and his blood pressure, blood sugar, or cholesterol is still high and the physician asks him whether he is taking his medication, studies show that patients tell their doctor they are taking the medication even if they are not in order to avoid an uncomfortable conversation.[3] The concerned physician then prescribes a second agent to treat the condition. If that is an expensive brand or nonformulary drug, the cycle repeats itself. We do not know what percentage of polypharmacy is caused by this dynamic, but it is likely to be considerable.

Properly implemented electronic prescribing (e-prescribing) could readily address this problem. For a patient with moderate hypercholesterolemia, the physician would open the e-prescribing system using the patient's demographic data, which would prompt the system to verify the patient's prescription drug benefit plan. The physician would then look up a specific drug or, even more appropriately, look up the condition to treat. For hypercholesterolemia, the e-prescribing system would ideally provide a list of therapeutic options and at the same time list the prices of all those options in terms of out-of-pocket cost to the patient and to the health plan, as illustrated in Table 18.1. The following are necessary to generate these data:

1. The e-prescribing system would have verified the patient's eligibility and the specific prescription drug plan. This is not complicated and is exactly what is done when a patient presents at a retail pharmacy and a pharmacy technician enters the patient's plan information into the pharmacy claims switch network.

2. The e-prescribing system would compare similar drugs on therapeutic equivalence. For example, Zocor is the brand version of simvastatin, so the two are dose equivalent. Lipitor at 20 mg and Crestor at 10 mg are therapeutically equivalent to simvastatin 40 mg. These data are available from work financed by the federal government and from comparative effectiveness analyses.[4]

3. The e-prescribing system would have priced the drugs based on the negotiated contract or mail-order rates from the patient's prescription drug plan.

In this instance, the e-prescribing system provides clinically relevant data on the pharmacotherapeutic options while doing the heavy lifting to calculate real drug costs based on complex and often arcane prescription drug plan specifics. For example, the patient's generic copay under this plan is $10 and his brand copay is $50 per 30-day supply. For generics like niacin, whose total price is less than $10, the patient pays the "lesser than" price—in this case, the total price of $7.89, which is less than the patient's $10 generic copay. Also, omega-3 fish oil is nonformulary and over the counter (OTC); under the patient's plan, this means that the patient pays the total cost of $15.39. Similarly, brand simvastatin (Zocor) and Crestor are nonformulary; according to this patient's prescription drug plan, this means that the patient pays the full price and the plan does not cover any of the cost.

Essentially, the data in Table 18.1, if presented to the physician at the point of care while he or she is prescribing, give the physician the ability to add cost easily to the therapeutic decision and facilitate a practical economic and financial discussion between the patient and the physician about what the best therapeutic option is relative to cost. In the end, we can hope that the outcome would be a lower cost prescription to treat the patient's hypercholesterolemia with higher compliance.

TABLE 18.1 Hypercholesterolemia Therapeutics: Example Costs for a 30-Day Supply

Drug	Formulary	Source	Total Cost	Patient Pays	Plan Pays
Niacin 1,000 mg	Yes	Generic	$7.89	$7.89	$0.00
Niaspan 1,000 mg	Yes	Brand	$126.23	$50.00	$76.23
Omega-3 fish oil 1,200 mg	No	OTC	$15.39	$15.39	$0.00
Simvastatin 40 mg	Yes	Generic	$6.70	$6.70	$0.00
Zocor 40 mg	No	Brand	$149.35	$149.35	$0.00
Lipitor 20 mg	Yes	Brand	$122.07	$50.00	$72.07
Crestor 10 mg	No	Brand	$111.87	$111.87	$0.00
Vytorin 10/40	Yes	Brand	$103.72	$50.00	$53.72
Zetia 10 mg	Yes	Brand	$97.34	$50.00	$47.34
Gemfibrozil 600 mg	Yes	Generic	$11.25	$10.00	$1.25

Some obstacles must be overcome to facilitate this scenario. Like the other aspects of our ideal healthcare information solutions, some of the obstacles are technical, some are market driven, and some are political.

The technical issues are the same as those we confront with EHRs: a current lack of true interoperability between information systems. For example, there is an NCPD standard for the exchange of formulary information, but that standard does not currently accommodate complex patient payment plan options or prescription drug pricing transparency. In addition, even standards-based data interchanges such as Surescripts for e-prescribing and RxHub for claims data—both managed by a not-for-profit organization owned by the retail pharmacies and pharmacy benefit managers—do not include all pharmacies or all pharmacy benefit plans.

The market-driven issues are that the retail pharmacies, both independents and chains, do not want prescription drug pricing transparency because, for all practical purposes, they are competing on service and convenience, rather than on price. Most patients on most prescription drug plans pay only their copay, so unless they are required to purchase through mail order or within a constrained pharmacy network, patients are just as blind as physicians to the overall drug cost and the cost to the health plan. Pharmacy benefit managers also do not want prescription drug pricing transparency because they do not want to reveal their negotiated discounts, either as pharmacy network discounts or as market share rebates, and they do not want to reveal their mail-order pricing.

The political issue is that the pharmaceutical manufacturers do not want prescription drug pricing transparency for competitive reasons. As illustrated in the previous pricing example, prescription drug pricing transparency would show the profound difference between the price for drugs with therapeutic equivalence such as generic simvastatin 40 mg at $6.70, Zocor 40 mg (brand simvastatin) at $149.35, and Lipitor 20 mg, still under patent at $122.07.

Note that if side effect profiles and efficacy were also provided to the physician by the e-prescribing system, niacin and omega-3 fish oil might be the safest combination to treat the patient, even if the incremental cost were somewhat higher than that of a statin like simvastatin. In instances like this, data on efficacy and side effects collected over time from large populations through structured EHRs with computable data illustrate the full circle that we could obtain collecting data and feeding it back into the care delivery process. This goes back to our ideal fundamental goal in healthcare information technology: to build a data flow continuum to improve clinical care and outcomes constantly.

18.5 CONCLUSION

Fully implementing healthcare informatics offers a number of promises for better patient care and lowered healthcare costs, but many issues must be resolved before this becomes a reality. These issues—technical, market based, and political—are not insurmountable, but they illustrate that constraints on healthcare information technology are complex and multifactorial. What we have to do, for the benefit of our patients, is to "grab the bull by the horns" and start addressing the constraints. As we strive to solve problems using information technology, we have to be diligent to use technology to simplify what we do, rather than further complicate it. In order to overcome constraints, we need to focus our efforts on the ideal goal of improving clinical outcomes and the efficient, affordable delivery of healthcare.

REFERENCES

1. Battleman, D. S., Callahan, M., and Thaler, H. T. Rapid antibiotic delivery and appropriate antibiotic selection reduce length of hospital stay of patients with community-acquired pneumonia: Link between quality of care and resource utilization. *Archives of Internal Medicine* 2002. 162:682–688.
2. Goldman, D. et al. Pharmacy benefits and the use of drugs by the chronically ill. *JAMA* 2004. 291:2344–2350.
3. Piette, J. et al. Cost-related medication underuse—Do patients with chronic illnesses tell their doctors? *Archives of Internal Medicine* 2004. 164:1749–1755.
4. Helfand, M., Carson, S., and Kelley, C. Drug class review on HMG-CoA reductase inhibitors (statins). Drug Effectiveness Project, Oregon Health & Science University, August 2006, Table 2, p. 12.

Index